Vanessa Farias Silva
Viviane Farias Silva
Carlos Vailan C.de Bezerra

Analysis of the efficiency of sustainable wastewater treatment

Vanessa Farias Silva
Viviane Farias Silva
Carlos Vailan C.de Bezerra

Analysis of the efficiency of sustainable wastewater treatment

Sustainable wastewater treatment makes it feasible to use it in different segments and on different scales

ScienciaScripts

Imprint

Cover image: www.ingimage.com

This book is a translation from the original published under ISBN 978-3-330-77210-6.

Publisher:
Sciencia Scripts
is a trademark of
Dodo Books Indian Ocean Ltd. and OmniScriptum S.R.L publishing group

120 High Road, East Finchley, London, N2 9ED, United Kingdom
Str. Armeneasca 28/1, office 1, Chisinau MD-2012, Republic of Moldova, Europe
Managing Directors: Ieva Konstantinova, Victoria Ursu
info@omniscriptum.com

Printed at: see last page
ISBN: 978-620-8-64092-7

SUMMARY

SUMMARY

The research was carried out with the aim of evaluating the efficiency of wastewater treatments. The experiment was carried out in three phases, delimited by the three hydraulic detention times, 7, 9 and 10 days, respectively. Analyses of the effluent from the UASB reactor and the FAC were carried out weekly, checking the removal of organic matter in the form of BOD5 and COD, solids and their fractions, nutrients (nitrogen and phosphorus) and indicators of fecal contamination (thermotolerant coliforms). The results showed that there were no significant differences in the FAC effluent between the three phases in terms of organic matter and solids (at the 5% level), but there were differences in terms of electrical conductivity, nitrogen and thermotolerant coliforms. The FAC performed well in terms of removing organic matter (65% for COD and 64% for BOD5 during phase 1[a]) and suspended solids (81% for TSS and 83% for VSS during phase F). Given these results, it can be seen that the FAC showed good efficiency in the post-treatment of anaerobic effluent, resulting in a clarified effluent with low concentrations of organic matter (12 mg BOD5.L^{-1}), suspended solids (<10 mg TSS. $L^{(-1)}$)) and turbidity (<10 UNT). With regard to sanitary quality, the effluent produced had thermotolerant coliform concentrations ranging from 10^3 to 10^4 CFU.100 $mL^{(-1)}$), therefore values above the standards set by the WHO (1989) for use in unrestricted irrigation.

Keywords: Post-treatment, Chicane filter, Anaerobic treatment.

1 INTRODUCTION

In Brazil, the lack of basic sanitation and adequate infrastructure has contributed to raw sewage being dumped in inappropriate places, such as rivers, lakes and dams. According to IBGE data (2000a), around 47.8% of Brazilian municipalities do not have a sewage collection system and of those that do, only 20.2% treat the waste before releasing it into the environment.

As a consequence of these discharges, some impacts are clearly observed, such as the compromise of public health through the contamination of water sources and the spread of waterborne infectious diseases, the interruption of agricultural or industrial activities due to the affected quality of the water, as well as other inconveniences such as the release of bad odors and fish mortality (IMHOFF, 1998).

These impacts result in two major problems that are naturally interlinked. The first concerns the continued contamination and pollution of surface water, compromising aquatic and terrestrial biodiversity. The second, which is just as important as the first, is related to the reduction in the availability of water resources, compromising the distribution of drinking water and meeting the need for an increase in its demand.

Even though the amount of water on the planet has remained practically unchanged since the dawn of time, the availability of water for public supply has been significantly reduced. This is due to factors such as: growing demand for consumption, pollution and enormous waste. The substantial increase in population implies a greater consumption of drinking water and, consequently, a greater amount of waste production which, more often than not, ends up in water sources. Another relevant factor is the enormous waste of drinking water, both in households and in water treatment plants and distribution networks.

In some towns, water not only meets basic needs, but also provides a strong boost to the economy and development of these regions. Therefore, the importance of guaranteeing the quality and quantity of water is far greater than those so commonly emphasized: animal watering and basic human activities (body grooming and general cleanliness). But it encourages and facilitates the establishment of industries and the breeding of animals; it even offers recreation for both adults and children, contributing to people's mental health.

Although Brazil holds a good part of the planet's fresh water, 12% of the reserves, the lack of proper management of water resources, such as springs contaminated by sewage or the waste of water during public supply, contributes to a significant reduction in drinking water for the population. Garcia (2003) reports that Brazil is in 50th place in relation to other countries when it comes to guaranteeing drinking water for the population. In order to make this claim, the author carried out a study which took into account five factors: the amount of fresh water per inhabitant; the share of the population with treated water and sewage; income, health, education and social inequality; water wastage; water pollution and environmental preservation.

This reality is attracting greater attention from researchers and mobilizing authorities. The International Water Forum held in the state of Rio Grande do Sul in 2003 brought together around 2,200 individuals and government representatives to discuss water resources: water supply and adequate sewage treatment; with the aim of guaranteeing access to drinking water, in quantity and quality, for all citizens (SILVA, 2003).

As is already happening in some parts of Brazil, the private sector is stepping in and taking over water supply and sewage treatment, as exemplified by the states of Sao Paulo and Rio de Janeiro.

In these states, the benefits of providing basic sanitation services have reached increasing numbers of people and the prospects for these companies are growing (VAZ, 2002). These benefits fall on the consumer at an additional cost, since private companies aim to make a financial return on their investments, which would not be the case if the state guaranteed the right to basic sanitation to the population.

Therefore, treating sewage is a proposal that is gaining momentum, as many of the springs that supply various regions have become large deposits of sewage, making the water totally unsuitable for any activity, requiring greater expenditure on its treatment in order to use it. And this proposal becomes more attractive when allied to the treatment is the use of the treated sewage; reducing the consumption of drinking water for less noble purposes, such as domestic cleaning or garden irrigation.

Domestic sewage treatment can be carried out using anaerobic processes, due to their many advantages over aerobic processes, such as reduced installation and operating costs, ease of operation and low waste production, making it possible to introduce them into poor communities. In anaerobic treatment using the UASB reactor, in addition to the advantages mentioned above, the effluent is retained for a short period of time, making it possible to treat larger volumes of sewage in a shorter space of time; this makes it preferable to other systems.

However, treatment via a UASB reactor does not offer total removal of organic matter, despite its excellent removal efficiency, nor does it offer the effluent sanitary quality. It is therefore necessary to combine the UASB reactor with another system that polishes the effluent so that it can be used in small communities or even discharged into the environment without harming it.

The use of a Chicana Anaerobic Filter for post-treatment of UASB reactor effluent is a technically and economically viable alternative, as it uses disposable materials such as PET bottles as filler, making the system less expensive and making it more difficult to clog the filter due to the high volume of empties.

2. OBJECTIVES

2.1.GENERAL OBJECTIVE

To evaluate the efficiency of treatment by UASB and a Compartmentalized Anaerobic Filter using PET bottles as a support medium for biofilm fixation in the post-treatment of sanitary sewage, enabling the generation of effluent within quality standards for reuse.

2.2.SPECIFIC OBJECTIVES

S To evaluate the characteristics of disposable polyethylene terephthalate as an alternative support material for biofilms.

S Check the FAC's performance over the three hydraulic detention times (7, 9 and 10 days).

S Evaluate the removal efficiency of organic matter, nutrients and indicators of fecal contamination.

S Quantify and characterize sludge production during the FAC operating period.

3. LITERATURE REVIEW

3.1SOCIETY: FROM SUBSISTENCE TO CAPITALISM

For thousands of years, human beings took from nature only what they needed to survive. This way of life was very favorable for two reasons: firstly, people had a large part of the day free and used it for parties, informal conversations, recreational and religious activities, in short, socially and physically healthy activities; secondly, there was no overexploitation of natural resources and everyone's survival was guaranteed for several generations. The economically active population in this way of life were the hunter-gatherers and they lived in groups with few resources, a fact that was attributed to their nomadic lifestyle, which prevented them from owning large and/or numerous possessions because it made pilgrimage difficult. Food was seen as a common good, they ate a nutritious and varied diet, they were social people, they performed rituals, weddings and other activities, they always sought the good of all the members of the group, everything was shared equally and they constantly met with other groups (PONTiNG, 1995).

With the development of technology and the formation of cities, a change in habits was inevitable. New perspectives, new desires and, therefore, new patterns of life were assumed by society, which went from nomadic to sedentary, from hunter-gatherer to farmer-producer, from subsistence to storage, making it greedy and selfish, an exclusionary and self-destructive consumer society. Ponting (1995) depicts how technological advances and society's incessant search for greater benefits have contributed to the devastation of natural resources and, to a greater extent, to the degradation of man himself, as a social being with intellectual capacities.

Today there is concern about the future results of human evolution. The Malthusian theory points to a global collapse with a lack of food for the future population, arguing that the planet will not be able to feed the population expected in a few years' time. Leonardi (2001) lists current global problems as symptoms of revolutionary transformations, citing ecological catastrophes, carbon dioxide emissions, the greenhouse effect, the desertification of fertile areas, the stockpiling of non-renewable energy reserves and others as worldwide problems that require immediate global solutions.

Lowy (1999) studied Marx's ideas and analyzed the conditions of production through the interests of the capitalists, since they are the ones who lead the current system that governs our society; coming to the conclusion that, according to its expansionist dynamics, capital is endangering, or even destroying, the very conditions for maintaining the capitalist system, starting with the natural environment, the supplier of all raw materials.

Pelizzoli (2002) assures us that the consequences of relations under capitalism are clear to see, and also quotes a statement by Sirio Velazco, in his book 'Ethics of Liberation', **"The current ecological crisis is inseparable from the alienated work in force under capitalism, work in which man alienates himself from himself to the extent and because he alienates himself from nature [...]".** This author believes that a new paradigm must be implemented. A paradigm that takes into account the economic needs of today's society, without forgetting ecological maintenance and the importance of other human values that are so often forgotten, such as equality, fraternity and humility among human beings, while always emphasizing the importance of other living beings, who are also indispensable in the natural cycle of life.

This concern has given rise to various proposals that have been discussed and defended in conferences and important documents. Bruseke (2001) describes the entire historical process surrounding the emergence of sustainable development. This theory was born out of ecodevelopment, a proposal by Canadian Maurice Strong in 1973, characterized by the concept of a new ecological-political-economic model centred on the interests of peripheral countries. Sustainable development took shape with the Brundtland report, in an even more complex vision, linking the economy, technology, society and politics, emphasizing a new ethical stance towards the responsibility of guaranteeing the survival of current generations and future generations.

Sustainable Development has emerged with the aim of restructuring, modernizing and innovating development models, so that economic growth means improvements in the standard of living of all segments of society and the maintenance of ecological balance, guaranteeing future prospects. Thus, Sustainable Development must establish strong pillars to break with the old paradigm established by capitalism, which is rooted in society's economy and culture. Leff (2001) draws attention to the fact that neoliberalism, or its followers, are trying to reverse or even misrepresent the practices and theories of Sustainable Development; the neoliberal proposal would be to value nature, putting a price on natural goods and services, arguing that this attitude would balance use and avoid ecological overload. The author considers the use of this proposal to be a precipitation to catastrophe, since trying to reduce a natural asset to a mere numerical value would be totally unreasonable, as would trying to put a value on the services provided by man and his potential, such as creativity, courage and even moral values.

Leff (2001) quotes an excerpt from a letter written by the chief of the Redskin Indians in response to the Great White Chief of Washington, when he made him an offer for the lands they occupied:

"How can you buy or sell the firmament or the warmth of the earth? If we don't own the freshness of the air or the shelter of the waters, how can you buy it? Every part of this land is sacred to my people. Every glistening pine forest, every grain of sand on the beaches, every drop of dew in the closed woods, every hill and even the sound of every insect is sacred to the memory and past of my people. The sap that circulates through the veins of the trees carries with it the memories of the redskins. We are part of the earth and it is part of us. The fragrant flowers are our brothers and sisters. The steep cliffs, the wet meadows, the warmth of the body of horse and man, we all belong to the same family (...) The crystal-clear water that flows in the rivers and streams is not simply water, but also represents the blood of our ancestors. The murmur of the water is the voice of my father's father (...) and every ghostly reflection in the clear waters of the lakes tells the facts and memories of the lives of our people (SEATLE, 1854)."

It is well known that the principle of sustainability is confronted with modernizing reason and the neoliberal economic process. It comes in response to a new rationality based on ecological potential and respect for cultural diversity. But the neoliberal proposal is strong, and is already taking on real proportions, as mentioned by Vaz (2002) with regard to private companies' efforts to take over basic sanitation. Letting them continue with this investment is to confirm the neoliberal proposal to appropriate natural resources through the acquisition of capital; it is to approve the paradigm that everything can be bought or sold, starting with such an important resource as water. Combating this initiative is the proposal to treat domestic sewage and use it in the community itself, which is an important alternative, especially if the state is ahead of the game and takes on this responsibility.

In addition, this proposal aims to reconcile urban growth with natural resources, in line with sustainability recommendations, because as well as treating sewage, the effluents are also used for non-potable purposes and are implemented in small communities.

It is based on the accumulative growth of socio-environmental problems that have been manifesting themselves with implications at various levels. Implementing it in a feasible way would meet the needs of our current society in terms of basic sanitation, especially in terms of sewage treatment; at the same time avoiding the pollution of water bodies that has been compromising public supply and the survival of many living beings, including humans.

3.2 SMALL COMMUNITIES

One of the biggest challenges faced by peripheral countries has been to implement and maintain projects that provide a basic sanitation infrastructure, especially when it comes to small or rural communities, because they are difficult to locate, dispersed or even have a low per capita income. In fact, these communities suffer the most from a lack of basic sanitation, and consequently have the highest mortality, birth and malnutrition rates.

For Ho (2003), poor sanitary conditions are not just a problem in rural areas, but many urban communities share a lack of infrastructure in terms of water supply and sewage collection. This is the result of a lack of urban planning, investment capital, skilled labor or even environmental education. Coupled with these situations are the high levels of disease and poverty that engender people without citizenship and an obsolete economy.

In order to reverse this situation, it is necessary to invest heavily in the implementation of sanitary sewage treatment, as there are high costs for appropriate pipes to collect and transport the sewage, qualified labor, installation and monitoring of the system, as well as the need to properly dispose of the effluent and waste produced, in order to provide these communities with the minimum sanitary conditions without harming the natural environment.

The costs of setting up a sewage treatment system can be significantly reduced, all that is needed is for the community to be willing to contribute to the project, and this contribution would not only be financial, but their active participation would be essential: in the administration, supervision and maintenance of the treatment system (SAUNDERS & WARFORD, 2002). Nevertheless, the community's acceptance and participation is essential for the continuity of a project of this nature, even though nowadays there is greater awareness and sensitivity on the part of the population with regard to the environment and issues of public interest (SANTOS & MANCUSO, 2003).

In northeastern Brazil, but specifically in the state of Paraiba, there are many municipalities with small populations. Table 1 shows the municipalities in Paraiba with a population of less than 10,000 inhabitants, resulting in more than 130 municipalities. Of these, 70 have a population of less than or equal to 5,000 inhabitants and around 30 have no more than 3,000 inhabitants.

It is clear that there are many small municipalities, in terms of population, that would probably embrace the proposal of alternatives for treating their sewage. Data from the IBGE (2000c) shows the restriction of basic sanitation for these municipalities, where, in the entire state of Paraiba, only 35% of households have all the basic sanitation lines, i.e. sewage, water supply and garbage collection, and 25% do not receive any basic sanitation service; for municipalities in Paraíba with a population of less than 10,000 inhabitants and more than 5,000, around 10% of households have adequate sanitation and 45% have a total lack of basic sanitation; municipalities with a population of less than 5,000 inhabitants are living a calamitous reality, as 52% of households are not served by basic sanitation services. Since the lack of conventional basic sanitation affects most of the population of Paraíba, this includes the absence of sewage disposal, causing sewage to be discharged in the open near homes, flowing into the springs whose waters are the only sources of supply, causing major and serious consequences; it is necessary to adopt alternatives for the treatment of this waste that alleviate this situation and that can be implemented with the utmost urgency, in order to provide the minimum sanitary conditions for these communities.

Table 1. Municipalities in the state of Paraíba with a population of less than 10,000 inhabitants.

Municipalities	Habit	Municipalities	Habit	Municipalities	Habit
White Water	8.377	Curral Velho	2.558	Pedro Régis	4.883
Aguiar	5.638	Damiao	3.645	Pylons	7.800
Alcantil	4.958	Desterro	7.701	Little pilots	5.430
Jandaira cotton	2.209	Diamond	6.920	Well of José de Moura	3.529
Amparo	1.886	Two Roads	3.818	Silver	3.425
Aparecida	5.894	Emus	3.061	Quixabà	1.308
Baraûnas Sand	2.104	Friar Martin	2.923	Riachao	2.793
Areial	6.039	Gado Bravo	8.521	Riachao do Bacamarte	3.948
Assumption	2.960	Gurjao	2.789	Riachao do Poço	3.694
Baia da Traiçao	6.483	Ibiara	6.383	Salty	2.823
Baraûna	3.169	Igaracy	6.307	Santa Cecilia	6.862
Barra de Santana	8.311	Itatuba	9.374	Santa Cruz	6.471
Barra de Sao Miguel	5.162	Jericho	7.416	St. Helena	6.170
Belém do Brejo do Cruz	7.011	Juarez Tàvora	7.081	Santa Inês	3.548
Bernardino Batista	2.559	Junco do Seridó	5.968	Saint Theresa	4.728
Boa Ventura	6.588	Juripiranga	9.647	Santana de Mangueira	5.773
Boa Vista	4.983	Lagoon	4.844	Santana dos Garrotes	7.882
Bom Jesus	2.193	Lagoa de Dentro	7.086	Santarém	2.568
Good Success	5.285	Ballast	3.118	Saint André	2.800
Bonito de Santa Fe	9.230	Livramento	7.605	San Francisco	3.464
Borborema	4.730	Street	3.389	Sao Joao do Cariri	4.703
Brejo dos Santos	5.948	Lucena	9.755	Sao Joao do Tigre	4.481

Cabaceiras	4.290	Mae d'Agua	3.459	Sao José da Lagoa Tapada	7.184
Indian waterfall	7.834	Malta	5.692	Sao José de Caiana	5.737
Cacimba de Areia	3.577	Marking	6.203	Sao José de Espinharas	5.109
Cacimbas	6.979	Marizópolis	5.618	Sao José de Princesa	4.970
Caiçara	7.325	Mataraca	5.500	Sao José do Sabugi	3.903
Cajazeirinhas	2.848	Matinhas	4.086	Sao José dos Cordeiros	4.136
Caldas Brandao	5.155	Mato Grosso	2.427	Sao José dos Ramos	4.900
Camalaù	5.516	Maturéia	5.032	Sao Mamede	8.018
Campo de Santana	9.388	Mounted	3.969	Sao Miguel de Taipu	6.086
Grass	4.180	Mount Horeb	4.112	Sao Sebastiao do Umbuzeiro	2.894
Caraùbas	3.401	Mulungu	9.189	Seridó	9.106
Carrapateira	2.160	Nazarezinho	7.272	Serra Grande	2.855
Casserengue	6.568	New Forest	9.421	Serra Redonda	7.307
Catingueira	4.748	Nova Olinda	6.457	Sawmill	6.678
Caturité	4.183	New Palmeira	3.573	Sertaozinho	3.444
County	6.495	Olho d'Agua	7.831	Sobrado	6.885
Congo	4.602	Olivedos	3.194	Peace and quiet	2.598
Coxixola	1.422	Old Gold	2.823	Triumph	9.053
Cubati	6.388	Parari	1.437	Umbuzeiro	9.192
Cuité de Mamanguape	6.124	Passage	1.979	Vàrzea	2.051
Cuitegi	7.254	Pedra Branca	3.692	Mountain View	3.121
Curral de Cima	5.323	Pedra Lavrada	6.617	Zabelê	1.853

Source: Adapted from IBGE, 2000b.

3.3 SANITATION IN BRAZIL

The relationship between basic sanitation and mortality rates is obvious, especially with regard to infant mortality. Children, the most sensitive segment of society, are considered the thermometer of quality of life. If the society in question is characterized by malnourished, abandoned and/or drug-dependent children, it is clear that the environment in which they live has factors that favor this human degradation. These factors are directly linked to the socio-environmental framework, unemployment, low schooling or even illiteracy and poor health conditions.

Studies carried out by Dr Sylvia Babu, a doctor who followed child development for ten years in six rural villages in Bangalore, India, found that around 40% of babies born during this period died before reaching the age of five; this was attributed to the consumption of contaminated water which caused acute diarrhea, dehydration, malnutrition and even death (RANDALL, 2003).

Table 2 shows the conditions of Brazilian households in terms of domestic sewage collection. For each region, the number of households with the type of drain or lack thereof and their respective percentage are listed.

Table 2. Demonstration of sewage disposal in Brazil and its regions.

BRAZILIAN HOUSEHOLD CONDITIONS

Region	Connection to the collection network	Septic tank	Another outlet	No outlets

BRAZIL	22 316	46,52	10 409	21,70	12 049	25,12	3190682	6,65%
SOUTH	1 977 231	25,37	3 706 685	47,55	1 932 225	24,79	178 529	2,29%
SOUTHEAST	16 142	73,50	2 663 693	12,13	2 858 962	13,02	294 729	1,34%
CENTER-	1 108 160	31,69	465 056	13,30	1 820 306	52,05	103 343	2,96%
NORTH	96 811	4,06%	1 280 051	53,70	870 958	36,54	135 551	5,69%
NORTH EAST	2 992 381	24,26	2 294 252	18,60	4567 116	37,03	2 478	20,10%

Source: Adapted from PNAD, 2003.

Table 2 shows that less than half of Brazilian households, 46.5%, are connected to a sewage system. Another 47% have some other type of connection and 7% of households dispose of their waste directly in ditches and streams. This is a very shameful reality, because it is known that even if domestic sewage receives some treatment in the household itself, there is the production of effluent that must be properly disposed of, that is, if there is no sewage collection system, this effluent will be released into the environment; adding to the amount of households that have no drainage system, there are more than 53% of Brazilian households that dispose of their sewage produced daily in open-air streams. In the northeast, the average is much higher, with more than 75% of households in this situation. Furthermore, these figures mask a more hostile reality, since many Brazilian municipalities only collect liquid waste and send it untreated to a nearby body of water, considering that the country has a high rate of sewage disposal without any kind of treatment.

You can compare the conditions of basic sanitation by zone in the Northeast of Brazil using the data shown in Table 3. When the situation of households is compared between urban and rural areas in the Northeast, the picture is even more deplorable. It can be seen in Table 3 that the rural area is characterized as a less favored sector when only 50,321 households in the Northeast, which corresponds to 1.5%, are connected to the sewage collection system. This is an everyday reality, where it is common to see sewage in the open in this region, forming small streams that end up in small lakes or reservoirs. These waters are used by the community itself, both for domestic purposes and for irrigating small crops, and are also used as recreational water, especially by children, who are once again victims of this neglect.

Table 3. Demonstration of sewage disposal by zone in the Northeast.

Condition of homes	URBAN		RURAL	
Collector network	2 942 060	32, 78%	50 321	1, 5 %
Septic tank	2 135518	23, 79%	158 734	4, 72 %
Other	3 143 102	35, 03%	1 424014	42, 39 %
No drainage	752 433	8, 38 %	1 725 613	51, 37 %

Source: Adapted from PNAD, 2003.

In view of this enormous sanitation deficit, together with the epidemiological situation and the socio-economic level of the population, it is necessary to acquire sewage treatment systems that can be applied to each specific population. And in particular, systems that are characterized by low cost and simplified operation, favoring sustainability as a whole. Consequently, the viability of the system under study is becoming ever stronger, as there is a growing need for its use in the face of the reality facing communities, especially the poorest ones.

3.4 liquid waste and its impacts

Throughout human history, there has always been the practice of dumping waste into nature; this was done without discrimination, either because of lack of information or because the consequences of such an attitude were not realized, since the population was small and their activities were rustic, allowing the affected natural part to be regenerated.

Nature has the characteristic of regeneration, but certain conditions must be met, such as a long period of time and favorable conditions. When a forest area is partially devastated, after some time it is reconstituted; if the population of a species is practically wiped out, its population returns to its original density or even higher, depending on the conditions; if a river has been contaminated or has suffered the discharge of polluting material, it tends to re-establish its initial quality standards, or even become an even more diverse river. However, this regeneration has some limitations, as happens when an area undergoes desertification or a river is considered 'dead' (VON SPERLING, 1996a).

Bodies of water recover by purely natural mechanisms; the discharge of pollutants reducing the concentration of dissolved oxygen in the aquatic environment, causing various disturbances and alterations in the natural biota, results in an imbalance in the ecosystem, and therefore compromises the quality of the water; but through physical, biological and chemical processes, the concentrations of pollutants are reduced and/or altered to the point of preventing their interference with the environment, making it possible to restructure the ecosystem (EIGER, 2003).

With the exorbitant increase in population density, this process is no longer feasible, even though water bodies have a very limited renewal capacity compared to their assimilation capacity (self-depuration). This means that water bodies cannot absorb the enormous amount of waste from human activities, thus causing serious environmental problems. It is necessary for the organic matter found in these discharges to be stabilized and the excess nutrients removed before being discharged into water sources, preventing natural biological processes from compromising water quality and aquatic biodiversity (CHENICHARO, 1997).

In addition to the ecological impacts, there is the risk of contamination by pathogenic organisms. Contaminated water is a major source of water-borne diseases and has been a major contributor to the mortality of children under the age of five (KELLER et al., 2004). There are many types of water-borne diseases, such as cholera and skin and eye infections, which can cause minor discomfort, temporary confinement and even death.

Liquid waste also causes other inconveniences, such as visual pollution and the emission of bad odors, causing discomfort to the local population. It also increases the risk of accidents, as has already happened in Paraiba - Brazil, with a disturbance in the distribution of piped water resulting in the formation of a sewage stream in the middle of the street in one of its cities, and in another it knocked down a primary school wall (RANDALL, 2003).

Although access to drinking water has increased greatly, this has not been reflected in the collection of the liquid waste produced, nor in its treatment and final disposal. When access to water was restricted to using a well or bucket access to a river, water consumption was much lower when compared to using public water supplies. As the increase in sewage production is proportional to water consumption, a major problem has arisen from the sanitary point of view, as the practice of dumping in ditches and on the ground has continued without discrimination. This reality is clearly evident on the outskirts of urban centers and can involve disastrous dimensions; when, for example, a peripheral population finds itself on the slopes of a hill, because in rainy seasons the ditches and small streams turn into torrents and all the waste is carried down the hill, causing major landslides, collapsing shacks and huts, affecting many lives (LOBO, 2003).

3.5 Sewage AND ITS CLASSIFICATION

Sewage is defined as liquid waste from human activity, whether industrial, domestic or commercial, as well as seepage, agricultural areas, surface water and rainwater (JORDÂO & PÊSSOA, 1995).

The composition of sewage, as well as its concentration and flow, varies greatly because it is influenced by the characteristics of the region. The quality and quantity of sewage produced by a given population is directly related to its socio-economic level, the region's climate, social and industrial development, and even the season of the year relative to the period in which the characterization is being carried out. Sewage of industrial origin has very specific characteristics, generally high hardness due to the presence of calcium and magnesium ions and/or high concentrations of suspended solids. Sanitary sewage also has its own characteristics, such as high concentrations of nutrients, high alkalinity and great danger in terms of the risk of transmitting diseases because it contains a wide variety of pathogenic organisms.

Metcalf & Eddy (2003) classify sewage according to the concentration of its constituents, distinguishing them as strong, medium or weak (Table 4).

Table 4. Classification of sewage according to concentration.

Components	Concentration (mg L)		
	Strong	Average	Weak
Total solids	1200	720	350
Total volatile solids	325	200	105
Total suspended solids	350	220	100
Volatile suspended solids	275	165	80
DBOf	400	220	110
COD	IOOO	500	250
Total nitrogen	85	40	20
Amomacal nitrogen	50	25	12
Total phosphorus	15	8	4
Organic phosphorus	5	3	1
Alkalinity (CaCOj)	200	100	50
Fat	150	100	50

Source: Adapted from Metcalf & Eddy (2003)

With regard to origin, von Sperling (1996a) classifies sewage into three groups:

1. Domestic sewage - from household, commercial or institutional activities.
2. Industrial sewage - predominantly from industrial activities.
3. Infiltration - from precipitation, urban drainage and leaks from the supply network.

The latter may not be characterized as sewage if it is collected in isolation. Its low concentrations of solids, turbidity and indicators of fecal contamination show that this liquid can be discharged without restriction into water sources or even used for unrestricted crop irrigation. In Brazil, the separator system for sanitary sewage is predominantly adopted, but it is known that there are many illegal connections to the rainwater network, compromising the quality of this liquid (Idem).

According to Jordao and Pêssoa (1995), sewage can be classified as follows:

- Sanitary sewage - consisting mainly of water from domestic activities (generated in bathrooms, laundries, kitchens, etc.), a small amount of rainwater run-off and possibly an insignificant amount of industrial waste.

- Industrial sewage - comes from any industrial use and acquires its own characteristics according to the process used.

3.6 SEWAGE TREATMENT

It is of great environmental and health interest that wastewater is properly treated before being released into the environment. Their treatment prevents environmental pollution, the spread of disease and provides a second source of water which, even if of inferior quality, can be used for non-potable purposes.

Sewage treatment carried out in sewage treatment plants is a reproduction of the mechanisms that occur naturally in nature. However, its controlled form allows the processes to take place in less time without causing inconvenience to the environment or the population (KATO et al., 1999). Treating sewage means removing and/or reducing the concentration of its constituents, such as: suspended solids; organic material (biodegradable); nutrients (mainly nitrogen, phosphorus and potassium) and pathogens. Treated sewage can be used for various purposes, including agriculture, as long as it complies with legislative standards (VAN HAANDEL & LETTINGA, 1994).

3.6.1 Biological processes

The wide variety of microorganisms present in sanitary sewage makes a major contribution to biological treatment. They are responsible for the biological removal of organic matter, using it as a source of energy through the oxidation-reduction process. This is a low-cost process because all the components needed for this process are already available in large quantities in the sewage itself.

Biochemical reactions, in addition to providing energy to microorganisms, promote the production of bacterial mass. The mechanisms responsible for this process take place through bacterial metabolism. The action of microorganisms in metabolism occurs in two distinct and simultaneous ways.

One, known as catabolism, refers to the use of organic matter exclusively to obtain energy and, as a result, the organic matter is stabilized by transforming it into mineralized products. Another term, anabolism, is the use of material to synthesize new cells, resulting in the production of more organic matter, which is then stabilized (VAN HAANDEL & MARAIS, 1999).

Metabolic activity takes place under two conditions. In the aerobic condition, the oxidation of organic matter occurs through the consumption of available oxygen. The bacteria use oxygen as an electron acceptor, producing water and carbon dioxide, which is why it is necessary to introduce oxygen into treatment systems such as activated sludge. In the anaerobic condition, which occurs in the absence of oxygen, the electron acceptor can be nitrate ($NO3^-$), sulphate ($SO4^{-2}$) or carbon dioxide (CO2) (VON SPERLING, 1996b).

In aerobic metabolism, the oxidation of organic matter requires less time than in anaerobic metabolism. However, around 70% of this process takes place through anabolism, and there is an intensified production of bacterial mass, which necessarily has to be properly treated afterwards because it is bioactive material. On the other hand, in anaerobic metabolism, the energy resulting from digestion is much lower; the bacteria tend to metabolize more slowly, intensely breaking down the molecules of the material, reducing it as much as possible, thus providing more energy to be used biologically and stabilizing the organic matter; the final product being mineralized solids (VAN HAANDEL & LETTINGA, 1994).

3.6.2 Anaerobic Digestion

Anaerobic treatment takes place when *there is* no oxidizing agent available in the environment to transform the organic matter, and the end products are gases, predominantly methane and carbon dioxide. This transformation takes place through a consortium of various anaerobic microorganisms and can be used both to avoid environmental pollution and to obtain alternative energy, since methane can be used for a variety of purposes (MOLETTA, 2005).

Anaerobic digestion is an ancient process used to stabilize sludge from the treatment of domestic sewage, and eventually industrial sewage. The bacteria responsible for converting organic matter into biogas belong to the methanogenic group, they have a very slow growth rate, and therefore their metabolism is usually considered to be limiting in the process (METCALF & EDDY, 2003). In this process, part of the energy obtained is converted into methane, which is released into the atmosphere or captured in order to be used for combustion.

The development and use of the anaerobic process was only possible after studies of biological processes and knowledge of the microbial biota involved, as well as chemical processes and their reactions. This process only occurs due to the presence of certain groups of microorganisms that work successively, promoting the conversion of complex, long-chain organic compounds into reduced substances, and require specific conditions to guarantee their own survival (FORESTI et al., 1999).

Van Haandel and Lettinga (1994) say that the anaerobic digestion of organic material, generally composed of: proteins, carbohydrates and lipids, goes through the following processes:

I- Hydrolysis - the organic material is solubilized and converted into simpler, water-soluble substances such as amino acids, sugars and fatty acids by exoenzymes that are excreted by the fermentative bacteria.

II- Acidification - the products derived from hydrolysis are converted into even simpler compounds (volatile acids, alcohols, aldehydes and gases: CO2, H2, NH3, and H2O) by bacteria that are only anaerobic. However, there is an important presence of facultative bacteria, which consume the traces of dissolved oxygen present in the medium, having to act quickly to prevent oxygen from spreading into the medium, which would be fatal for the anaerobes.

III- Acetogenesis - acetogenic bacteria use the by-products of acidification and transform them mainly into acetic acid, hydrogen and carbon dioxide.

IV- Methanogenesis - stage characterized by the removal of organic material or its digestion, the products of acid fermentation are converted into stable gaseous products: methane and carbon dioxide.

In the same way, Figure 1 shows the diagram of the decomposition of organic matter by the anaerobic process, going through the four stages mentioned above, up to the formation of methane gas.

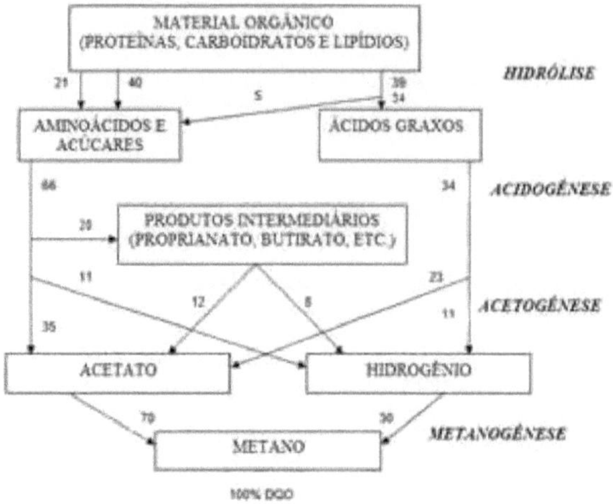

(VAN HAANDEL & LETTINGA, 1994)

Figure 1 Diagram of the organic material conversion process in anaerobic digestion.

3.6.3 Intervening Environmental Factors

For an efficient system, there needs to be a bacterial population that is adapted to the type of organic material and its effective load. This population needs favorable environmental conditions for its survival, such as: temperature, pH and absence of toxic substances.

Van Haandel and Lettinga (1994) describe how these factors influence bacterial metabolism and argue that it is important to observe them when applying anaerobic digestion:

- Temperature - a factor of great importance in anaerobic digestion, especially with regard to microorganisms. Each type of bacteria has its own optimum growth temperature, in the case of the bacteria responsible for the anaerobic digestion process, the optimum temperature can vary between 30 and 40°C.

- pH - methanogenic bacteria, the main culprits in the process of anaerobic degradation of organic matter, maintain normal activity when the environment has a pH within neutrality (6.0< pH< 7.8). During acid fermentation there is a predominant production of acetic acid which tends to lower the pH of the reactor, inactivating the methanogenic bacteria and causing an imbalance, which can lead to a collapse in the system. This happens if there is no balance between the fermentative and methanogenic bacteria. In the case of sanitary sewage, this phenomenon does not occur due to the significant presence of the carbon system (H2CO3, $HCO3^-$, $CO3^{-2}$), which is responsible for pH stability in the system.

• Toxic substances - chemical substances such as heavy metals and organo-chlorine substances are toxic to methanogenic bacteria, even in small concentrations; however, in the case of domestic sewage, the presence of these substances is not detected. However, there is the presence of sulphide, generated in the reactor from sulphate, which probably doesn't reach a high enough concentration to interfere. Another toxic substance is oxygen, which is found in sewage, but is removed by facultative acidogenic bacteria when in small quantities.

3.6.4 Intervening Factors in Metabolic Activity

In addition to the environmental factors, such as pH, temperature and the absence of toxic substances described above, which influence bacterial growth and metabolism, Chernicharo (1997) mentions two other important factors which are related to biological activity in terms of the removal of organic material: cell retention time (CRT), which is the time the solids remain inside the system, also known as the age of the sludge, and hydraulic detention time (HDT). Of the two factors within a project, the TDH is the most important because it is the simplest to specify and because the TRC depends on it.

The solids need to remain in the system long enough for the metabolic process to be completed and for the biomass to be treated and stabilized. Before the solids can penetrate the biofilm, they need to be hydrolyzed so that they can be absorbed and metabolized. When it comes to biofilm, hydrolysis, the initial phase of anaerobic digestion, takes place in a liquid medium outside the biofilm; Janning et al. (1997) observed, through experiments with biofilm under anaerobic conditions, a high release of soluble TOC indicating the hydrolysis of organic particles in a liquid medium.

3.6.5 Advantages of the Anaerobic Process

Under natural conditions, anaerobic decomposition requires three times as much time as aerobic decomposition, but aerobic decomposition comes at a high cost, especially in terms of energy consumption and the use of sophisticated devices needed for proper flow distribution and aeration, making treatment expensive (CHERNICHARO, 2001).

Operating costs in the anaerobic process are much lower, as well as the removal of organic material being more efficient, since complete oxidation is promoted and, therefore, sludge is produced in smaller quantities and is more stable when compared to the aerobic process. In addition, studies into new technologies have contributed to improving the reaction speed in the anaerobic process, resulting in a significant reduction in the time required for this process

(FORESTI et al., 1999).

Another great advantage of using anaerobic systems is the energy potential obtained from the combustion of methane produced during digestion. During anaerobic digestion, organic matter is converted into methane, a substance that releases 2.98 $Kcal.g^{-1}$ COD during combustion (VAN HAANDEL & LETTINGA, 1994).

3.7 ANAEROBIC FILTER

3.7.1 Biological filter

Knowledge of anaerobic filter technology began to be disseminated in 1969 in a publication that showed an efficiency of 80% in COD removal in a small TDH. A decade later, the implementation of this filter began in Brazil, with significant performance in the treatment of industrial sewage (ANDRADE NETO et al., 1999). ABNT (NBR 7229/82) standardized the filter 'model', strengthening and spreading this new technology.

The treatment of sewage by means of a biological filter occurs through the constant passage of affluent into the interstices of the support medium, promoting the growth and adherence of a biological mass on its surface, forming a film of bacteria, which absorbs a quantity of organic matter and digests it (KATO et al., 1999).

3.7.21 mplications of Flow Direction on Anaerobic Filter Efficiency

Anaerobic filters can have their flow in three directions: a) upward, b) downward or c) horizontal, as can be seen in Figure 2. Of these three, the upward flow is the one that has been applied the most in experiments, but rarely on a real scale. Some research is currently being carried out and applied to downward-flow anaerobic filters, checking their efficiency and level of operational ease.

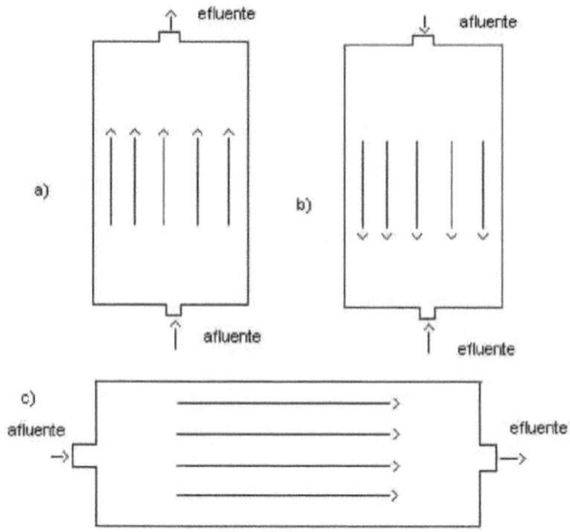

Source: Author

Figura 2. Direction of flow. a) upward b) downward c) horizontal.

The direction of flow influences the retention and sludge formation characteristics. In upward flow, the concentration of sludge is high, characterizing good efficiency and low drag of solids in the effluent, consequently increasing the risk of clogging. When the flow is downward, working in a completely drowned manner, the sewage has a better distribution resulting in greater efficiency and easier removal of excess sludge, reducing the risk of clogging (ANDRADE NETO et al., 1999).

Through comparative experiments, Andrade Neto et al. (2001) investigated the performance of two anaerobic filters that had the same configuration and the same support material, differing only in the direction of flow, one ascending and the other descending. He concluded that both filters operated under the same conditions had similar performance.

Therefore, the choice of flow used in the anaerobic filter must depend on the operational facilities and the possibility of reducing investment. In the obvious case, both the upflow and the downflow should operate with the bed completely drowned, i.e. with all the support material submerged by the effluent.

3.7.3 Chicane filter

The baffle biological filter is a pre-dimensioned container with chambers arranged horizontally in series, each separated by vertical walls or baffles. The sewage flow in each chamber is vertical and upward, as shown in Figure 3, and the concentration of organic matter decreases as it progresses through the chambers. The bed is fixed and made of crushed stone or other inert material (KATO et al., 1999).

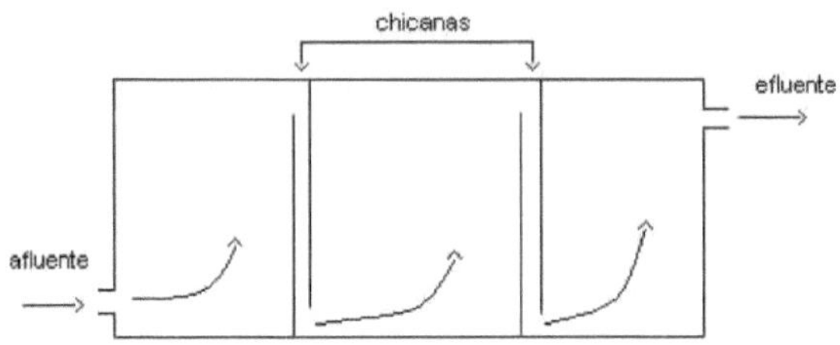

Source: Author

Figura 3. Schematic representation of a baffle filter.

The baffles function as a device that forces the liquid to make several upward movements, covering a greater area of the container; promoting a significant reduction in dead zones and preferential paths, factors that affect the performance of the system by reducing the useful volume used in the treatment. It can be considered that the greater the number of baffles, the smaller the reduction in these factors will be; however, too many can hinder the passage of the liquid and even cause it to become obstructed.

3.7.4 Relevance of the Post-Treatment Filter

The anaerobic filter can be used in the primary treatment of sewage, as it has many advantages over anaerobic reactors: it is resistant to variations in the effluent and has good stability, it does not need inoculum for start-up, it offers enormous design freedom, and it is very simple to build and operate. However, they are more suitable for post-treatment of other anaerobic units or other processes, due to the high risk of clogging when the effluent has high concentrations of suspended solids (CHERNICHARO, 2001).

Kato et al. (1999) also makes the same suggestion, however he argues that several factors, in addition to the high solids content in the effluent, make it possible for the filter to clog or clog, such as excessive growth or poor distribution of the biomass in the system, which could jeopardize the treatment.

The coupling of a UASB reactor with a biological filter can be used to treat sanitary sewage in tropical and subtropical areas with great success, resulting in low production of stabilized sludge and removal of more than 80% of COD and suspended solids (KELLER et al., 2004).

3.7.5 Aggregation of solids in the filter

The retention of organic matter contributes to sewage treatment by speeding up digestion. The biomass formed when stabilized and adapted to the sewage makes it possible for a greater organic load to be introduced into the system while maintaining good efficiency, resulting in the treatment of a greater volume of sewage in a shorter time (KATO et al., 1999).

The development of technology using adhered biomass is growing every year. This technology has met the need to acquire a system that provides a reduction in project area and reactor volume, reduces investment costs and provides effluent polishing; since this process simultaneously removes organic matter, nutrients and suspended solids (FARABEGOLI et al., 2003).

Mechanisms for immobilizing biomass are a way of increasing the time that solids can remain in a system, without having to increase the TDH; allowing the biomass to remain for the minimum time necessary to multiply and promote biological reactions, even when operating at a low TDH (CHERNICHARO, 1997).

The organic matter retained in the anaerobic filter can be found in three forms:

1. Flocs - made up of organic matter and microorganisms, they are agglomerated to form a somewhat complex structural unit (Figure 4a).

2. Dispersed biomass - similar to flocs, but dispersed in the liquid and retained in the interstices of the support material and less complex (Figure 4b).

3. Biofilm - consisting of a thin film adhered to the support material made up of organic matter and microorganisms (Figure 4c).

Figure 4 shows the three ways in which biomass is retained in the treatment system.

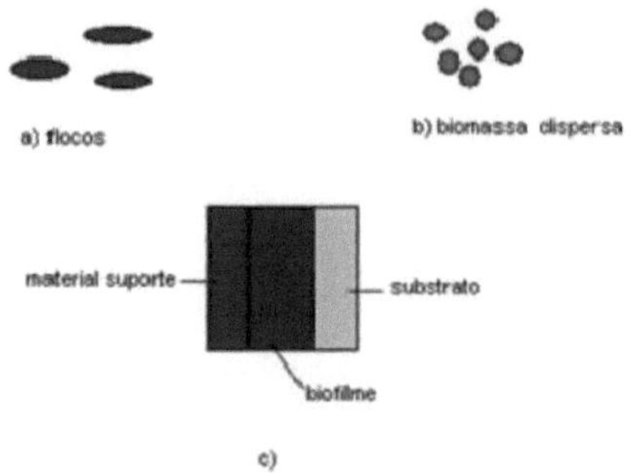

Source: author

Figura 4. Forms of biomass immobilization.

Biological treatment is influenced by the predominance of one of these forms, with biofilm formation being the one that most characterizes the filter and is therefore the most responsible for removing organic material.

3.8 SUPPORT MATERIAL

When constructing a biological filter, the choice of support material is essential, as it must meet the needs of the project and suit the characteristics of the liquid to be introduced. Kato et al. (1999) describes the characteristics that these materials must have:

- Resistant and light enough;
- Biologically and chemically inert;
- Facilitate flow distribution and make obstruction difficult;
- Low price;
- Easy to acquire.

The type of material used, its particle size and its specific area have a great influence on the filter's performance, and are directly related to its operation. The support material helps to promote greater interaction between the organic matter in the effluent, the biofilm and the solids in the interstices through more uniform flow, and ultimately acts as a physical barrier acting as a filtering process (ANDRADE NETO et al., 1999). In addition to these purposes, the choice of support material influences the adherence and growth of the biofilm and should promote uniform microbial distribution. It can be made of different materials and have various shapes (CHARACKLIS & MARSHALL, 1989).

However, it is utopian to believe that you can buy a material that fits all these characteristics. However, it is necessary that it has the minimum characteristics required to be used, meeting the expectations of the project.

In Brazil, the most commonly used material is crushed stone No. 4, which, in addition to being a heavy and expensive material, has a low void ratio (50%), which has implications for volume and the ability to accumulate active sludge (CHERNICHARO, 2001).

Andrade Neto (1997) mentions alternative support materials used by researchers in successful experiments, such as bamboo rings and waste from the steel industry (blast furnace slag); this substitution was intended to use material available in the region, with the aim of reducing operating costs and making the system more applicable.

3.8 BIOFILM

The adherence of microorganisms to any inert material used as a support in an aquatic environment forms a biofilm. Attachment depends on the natural tendency of microorganisms to adsorb to the surface and reproduce, along with the physical and biological processes that occur simultaneously. The biofilm is an extremely heterogeneous system; it is made up of a biotic fraction, which in addition to microorganisms such as bacteria, fungi and protozoa also includes macroorganisms such as worms and larvae; and an abiotic fraction, such as gases and degradable or inert substances. Generally, the adherence of microorganisms is related to the physical factors of sorption: absorption and adsorption (CHARACKLIS & MARSHALL, 1989).

The biofilm, seen microscopically, has several channels that allow the exchange of substances: substrate in and stable products out, with a constant exchange with the external and internal environment of the biofilm (KATO et al., 1999). Transport phenomena in the biofilm are relatively constant, substances penetrate and are released through the biofilm pores or temporary cracks and fissures that facilitate this passage (REICHERT & WANNER, 1997).

3.8.1 Advantages and disadvantages

In the treatment of sanitary sewage, the use of biofilms has a number of advantages, including the following (CHARACKLIS & MARSHALL, 1989):

-Ease of operation of the treatment system.

-Low cost.

For Andreottola et al. (2005) biofilm attachment is an alternative for sewage treatment characterized by high organic load removal due to the aggregation of organic matter in the support material, among the advantages of this alternative can be included the small area used for the project, making it unnecessary to recirculate the effluent and wash the filter.

Lazarova and Manem (1995) cite other advantages of using fixed biomass processes when compared to conventional biological treatments: rapid entry into the regime, high volumetric removal rate and increased process stability. Bishop (1997) includes as an advantage the wide variety of microbial populations contained in the biofilm, which enables the degradation of different substrates, as well as the presence of glycocalyx around the microorganisms, which protects them from the action of toxic agents.

Another advantage is the high concentration of biomass, which ensures a longer sludge age, resulting in lower sludge production.

The main disadvantage is the lack of control over the growth of the biofilm, causing the filter to clog. During the anaerobic activity of the biofilm, gases (H2S) are produced which, when released from the biological mass, make it susceptible to detachment from the support material. The increase in the thickness of the biofilm, the release of gases together with the hydraulic loads cause the phenomenon of "slumping", forming sludge that can be removed by sedimentation (JORDÂO & PESSOA, 1995).

The hydraulic loads that help the biofilm to detach from the support material are related to the shear stress forces. The

The thicker the biofilm, the smaller the empty spaces between the support material through which the fluid flows; consequently, the greater the force of the fluid and the easier it will be to carry the biological mass.

3.8.2 Adsorption

The adsorption process is based on interactive forces of attraction, in which the molecules of the substance in a liquid medium occupy vacant active sites on the surface of another solid substance, adhering or even attaching themselves to it (ALMEIDA, 2004).

Sewage treatment using adsorption has traditionally been used in industrial water treatment through the use of activated carbon to polish effluent after conventional biological treatment, removing the remaining organic matter. The authors attribute the phenomenon of adsorption to chemical and physical reactions, stating that when adsorption occurs by chemical processes, the bonding of the molecules is very strong and generally irreversible; but when adsorption is the result of physical phenomena, the bonds are formed by Van der Walls forces and the molecules are easily removed because they are weak bonds, and therefore a reversible process. (METCALF & EDDY, 2003).

Adsorption is responsible for the initial formation of the biofilm, since the adherence of the organic material to the support medium begins due to the predominance of this mechanism, and then interacts with the biological processes. As we have seen, adsorption can be reversible or irreversible. When it's reversible, the material is easily detached; when it's not, the material adheres to the support medium, allowing for greater retention of the material, initiating the formation of the biofilm. Together with the adherence of the organic material, there is the adherence of the microorganisms, which find food and fixation as a means conducive to their survival (CHARACKLIS & MARSHALL, 1989).

3.8.3 Chemical and Physical Properties

The chemical, physical and biological properties of the biofilm depend on the type of support material. The predominance of organisms modifying the microenvironment is a way of modifying their own metabolic activity.

According to Characklis and Marshall (1989), biofilms contain a lot of water, around 87 to 99%, and are generally very hydrophilic; another physical characteristic is their surface density, calculated as the mass per area of the substrate divided by the thickness of the biofilm. The thickness of the biofilm is a highly variable parameter due to its morphological characteristics and is a function of the age of the biofilm.

$$\rho_f = \frac{X_f}{L_f} \qquad (1)$$

Where:

ρ_f: density (g/cm^2)

x_f: concentration of the biofilm area (g/cm^3)

L_f: thickness (cm)

3.8.4 Biofilm Growth Rate

Both physical, chemical and biological transformations take place over a period of time. The efficiency of organic matter removal in the biological treatment of sewage also depends on the hydraulic retention time of the reactor. The time required for specific changes to occur is inversely proportional to the rate at which the process occurs.

The accumulation rate is a very important quantity because it can be used to correlate other factors involved in the process: temperature, concentration, fluid velocity and even the geometry of the support material. It also describes the speed of various phenomena such as chemical reactions and microbial cell adsorption.

Applying a material balance to the reactor substrate, we obtain the following equation:

Where:

$$\frac{VdC}{dt} = Q(C_i - C) + Vr \qquad (2)$$

C: concentration in the reactor (Kg/m^3)

Ci: feed concentration (Kg/m^3) Q: flow rate (m^3/h)

r: production rate per unit volume (Kg/m^3.h);

V: reactor volume (m^3)

Assuming that the concentration does not vary with time $\left(\frac{dC}{dt} = 0\right)$ then the rate of transport *dt* is equal to the negative rate of production:

$$D(C_i - C) = -r \qquad (3)$$

Where D is the dilution rate: $D = \frac{Q}{V}\ (h^{-1})$ (4)

The production rate (r) can be calculated using the reactor volume:

$$r = \frac{R}{V} \tag{5}$$

Where:

V: reactor volume (m3)

A: conversion rate (Kg/h)

The conversion rate can be based on the biomass and the specific reaction rate term:

$$r_X = \frac{R}{x} \tag{6}$$

Where:

x: biomass (Kg)

rx: reaction rate per unit of biomass (h^{-1})

4. MATERIAL AND METHODS

4.1 INSTALLATION

The pilot-scale experimental sewage treatment system was installed and kept in operation in an area belonging to the Companhia de Agua e Esgoto da Paraiba (CAGEPA), located in the Tambor neighborhood in the municipality of Campina Grande -PB (south latitude 7° 13', west longitude 35°52', and 550m altitude), where the Experimental Station for the Biological Treatment of Sewage (EXTRABES) and the laboratory of the basic sanitation research group (PROSAB) are located.

4.2 OPERATING SYSTEM

The system consisted of a UASB reactor made of fiberglass with a capacity of 5 m^3 and an anaerobic biological filter, also made of fiberglass with chicane partitions inside and an average capacity of 1 m3. Figure 5 shows a schematic of the system, made up of all the components, as well as the sequence in which the sewage passes through to the final treatment (FAC).

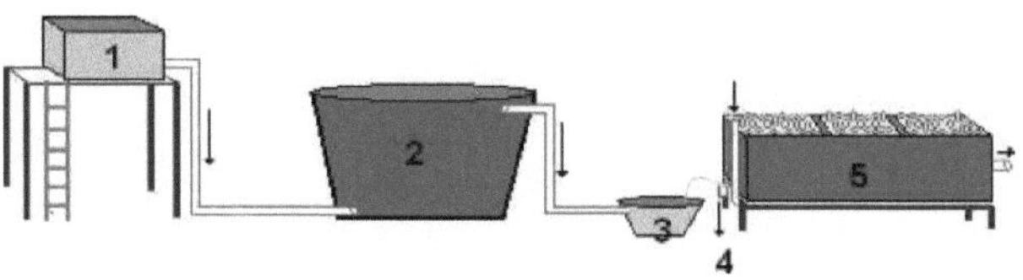

1-Equalization tank

2- UASB

3-Feed tank

4- Pump

5-FAC

Source: author

Figure 5 Schematic of the treatment system.

4.3 OPERATIONAL CHARACTERISTICS OF THE SYSTEM

4.3.1 UASB reactor

A portion of the sanitary sewage produced in the urban area of Campina Grande was collected and sent to an equalization tank in order to control the flow into the reactor. The sandbox itself is located 2.5m above the ground. This unit is part of the pre-treatment and is of great importance and even indispensable as it guarantees the continuity of the treatment, retaining coarse material such as wood, plastic and sand.

The effluent was conveyed by gravity to the UASB reactor at a flow rate of 20 $m^3.day^{-1}$, where it received the first anaerobic biological treatment. The UASB effluent was then fed to the FAC via a pump that was constantly calibrated to maintain the desired flow rate. Figure 6 shows a UASB reactor similar to the one used in the experiment, showing its interior divisions, the three-phase separator and the gas baffles.

(a) 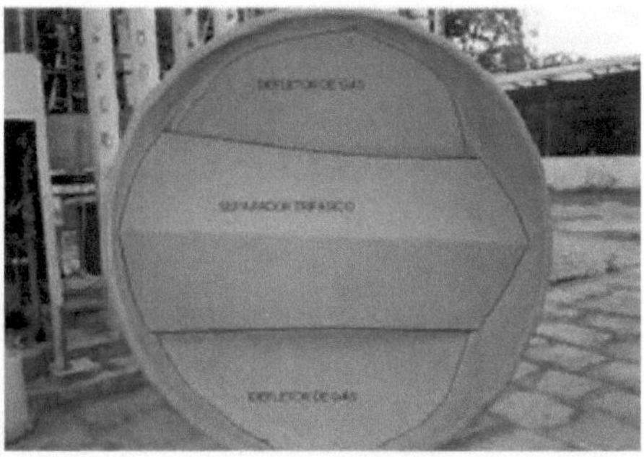

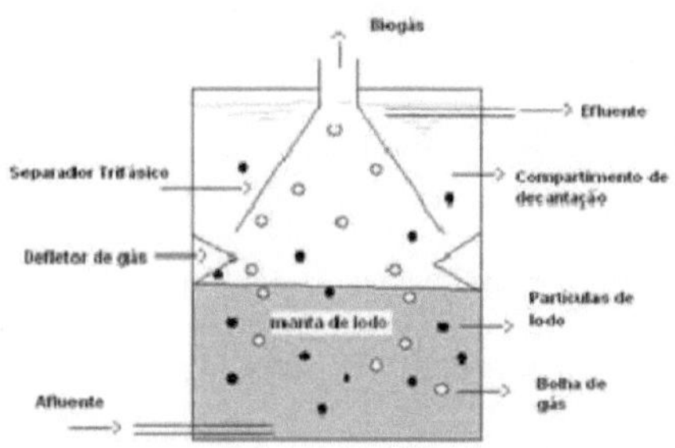

(b)

Source: author

Figura 6. UASB and its internal partitions: a) Top view; b) Internal operating diagram.

The main physical and operational characteristics of the UASB reactor are shown in Table 5.

Table 5. Physical and operational characteristics of the UASB reactor.

Features	Value
TDH (h)	6

Volume (m^3)	5
Height (m)	1,5
Diameter (m)	3
Specific organic load (kg $COD.m^{-3}.day^{-1}$)	2,52
Dispersion number(')	0,08
Dead volume fraction(') (%)	7,55

Source: Lima (2001).

4.3.2 Anaerobic Chicane Filter (ACF)

In the FAC, the effluent from the UASB reactor was fed from the bottom of the filter, always maintaining an upward flow in each chamber, justifying the decreasing height of the baffles. According to Figure 7, the filter had three chambers placed in order to improve the hydrodynamic performance of the filter by helping to reduce preferential paths and dead volume, so that the effluent remains in the system for the time theoretically estimated.

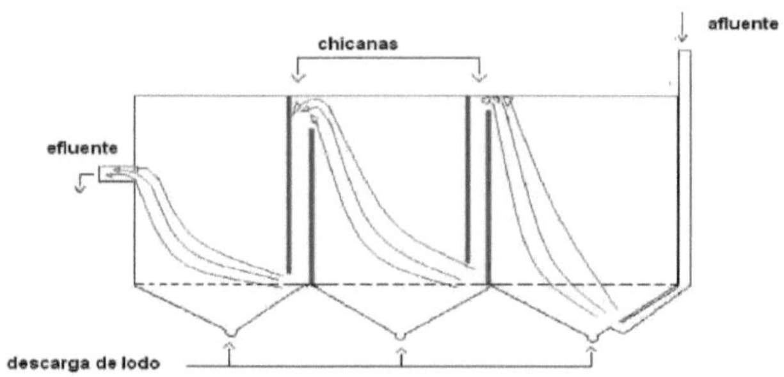

Source: author

Figura 7. Schematic of the baffle filter and the upward movement of the fluid

Figure 8 shows how the FAC works and how it is filled with PET bottles. Before the effluent enters the FAC, there is a feed tank for which the UASB effluent was destined in order to maintain control of the FAC feed. The configuration of the FAC, as well as its entire design, can be seen in Figure 9.

Source: author

Figura 8. Operation of the FAC.

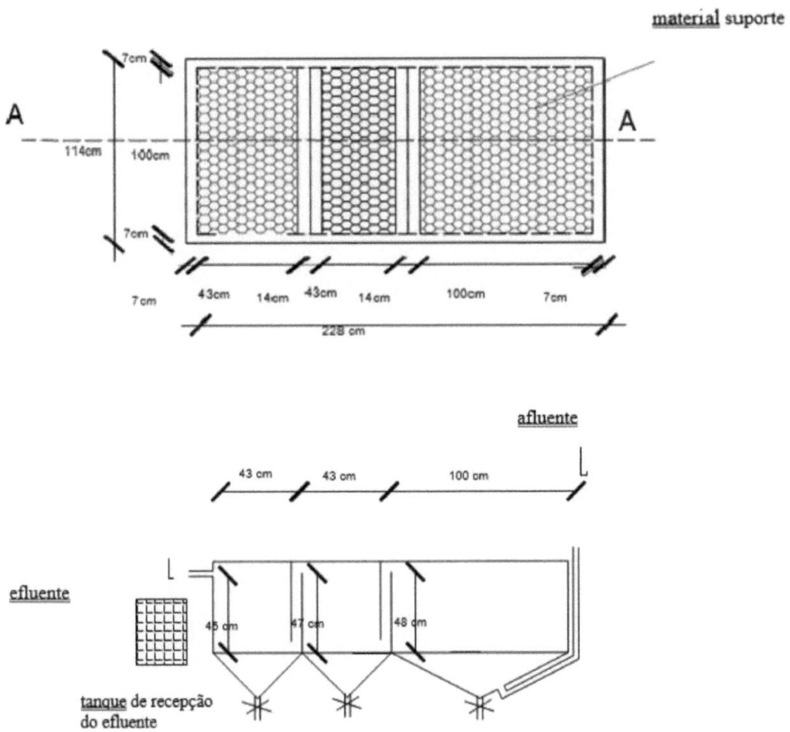

Source: Author

Figure 9. Dimensions of the FAC.

4.4 SUPPORT MATERIAL

The material used as a support for biofilm adhesion was a two-liter disposable bottle (polyethylene terephthalate - PET). The bottles were cut twice across to form rings (Figure 10).

Source: author

Figure 10. Material used as a support medium in the FAC

The PET bottles were obtained in the city itself from collectors and reused in the experiment. The void volume was determined by the difference in liquid displacement, using a 30-liter container. The specific surface area of the support material was determined experimentally through measurements.

In addition to the high void content, the use of this material has several other advantages:

- Easy to acquire and low cost;
- Inert material;
- Contributed to material recycling;
- Light and resistant;

4.4 OPERATING SYSTEM MONITORING

The experiment lasted ten months, divided into three phases. The first phase lasted four months, the next two three months each; each with a different hydraulic detention time (HDT) and flow rate, as shown in Table 6.

Table 6. Characteristics of each stage of the experiment.

Stages	Period	Flow (L/h)
1st Phase	August to November	5,932
2nd Phase	December to February	4,614
Stage 3	March to June	4,152

Source: author

The effluent from the UASB reactor was collected at the inlet of the FAC, and the second effluent at the outlet of the FAC. Amber glass containers were used to collect the effluent, which was cleaned and dried, and immediately transported to the laboratory (located 15 meters from the system) where all the physico-chemical and microbiological analyses were carried out.

The system was monitored daily, checking the flow rate, the operation of the pump and its proper cleaning, and the daily feeding and cleaning of the feed tank. The sandbox was cleaned from time to time, removing excess sand and any matter that had settled in it, ensuring that the tank had a 20cm layer of sand.

4.5 SLUDGE PRODUCTION

Three attempts were made to dispose of the sludge during the experiment, as shown in Table 7. The aim of these disposals was to reduce the risk of clogging and to encourage the effluent to circulate in the filter. Total solids analyses were carried out on the sludge.

On the first disposal, around 3L was removed from each chamber. No sludge was observed in the next two, but 3L was also removed from each effluent chamber as a precautionary measure.

Table 7. Sludge disposal from the FAC.

Disposal	Period	Situation
1°	October 2004	Presence of sludge
2°	April 2005	No sludge
3°	June 2005	No sludge

Source: author

As shown in Table 7, only the first situation is sludge disposal, as it was the only one in which there really was sludge. The other situations, even though they did not have sludge in their disposal, were included in order to clarify and even prove that the three attempts were made.

5. ANALYTICAL PROCEDURES

All the analyses carried out complied with the requirements described in APHA (1995). Table 8 shows all the physical and chemical analyses carried out in the laboratory on the effluents under study, as well as the procedures and equipment used.

Table 8. Physical and chemical analyses carried out during the experiment.

Analysis	Method	Equipment
Ammoniacal nitrogen	Semi-micro Kjeldhal method	Tecnal distiller model TE-036/1
Total nitrogen	Semi-micro Kjeldhal method with digestion.	Tecnal distiller model TE-036/1, Tecnal digester model TE-007
Total Phosphorus	Spectrophotometric with ascorbic acid and persulphate digestion	Milton Roy spectrophotometer model LR-45227, Phoenix autoclave model AVI 8
Soluble orthophosphate	Spectrophotometric with ascorbic acid.	Milton Roy spectrophotometer model LR-45227
Biochemical Oxygen Demand (BOD)	Direct reading	Ysi oxygen meter, model 54A
Chemical Oxygen (COD)	Dichromate oxidation-reduction titrimetry.	Jundilab digester model PN 456
Total Solids	Gravimetric	Fanen oven model 62700, Sartorius analytical balance
Total Volatile Solids	Gravimetric	Furnace model 62700, Fanen oven model 62700, Sartorius analytical balance
Suspended Solids	Gravimetric	Quimis water bath, Fanen oven model 62700, Sartorius analytical balance
Suspended Solids Volateis	Gravimetric	Furnace model 62700, Sartorius analytical balance.
Electrical conductivity	Instrumental method (direct reading).	Yellow Spring model M-33 conductivity meter
Turbidity	Instrumental method (direct reading)	Orbeco Hellige turbidimeter model 966
pH	Potentiometric	Orion pH meter model 230A

5.1 ALKALINITY

Alkalinity is the amount of ions present in a liquid that neutralize hydrogen ions. Most of these ions are bicarbonates (HCO3-), which is why they are most responsible for neutralization, but in addition to these there are other groups of ions, such as: OH^-, $CO3^-2$ (VAN HAANDEL & LETTINGA, 1994).

The use of anaerobic processes in the treatment of sanitary sewage has a great advantage in terms of its high buffering power. This characteristic is attributed to high alkalinity values, resulting from supply, rainwater runoff and adsorption during domestic use (METCALF & EDDY, 2003).

To quantify this parameter, the Kapp method was used, which consists of acidimetric titration up to three pH values (5.0, 4.3 and 4.0) using a strong acid (hydrochloric acid) of known concentration, using the following equations to obtain alkalinity total and bicarbonates, and the concentration of VFA:

$$AT(mgCaCO3/L) = Vac(4,3)x50.000xN/Vam \quad (7)$$

$$AB(mgCaCO3/L)= -(1312,26xNacxVac(5-4)/Vam)+0,0206AT+0,11 \quad (8)$$

$$VFA(mgHAc/L) = (131340xNacxVac(5-4)/Vam)-(0.0616\ AT)-10.9 \quad (9)$$

In which:

Vam - sample volume, usually 50 mL;

N - acid normality in eq/L;

Vac(5-4) - volume in mL of acid used to titrate from pH=5 to pH=4

Vac(4.3) - volume in mL spent to titrate the sample from initial pH to pH=4.3

5.2 MICROBIOLOGICAL TESTS

Characterizing effluents in terms of the presence of fecal contamination indicators is one of the essential requirements for sanitary sewage treatment. Bringing effluent within the norms of environmental legislation requires the determination of the thermotolerant coliform group.

Although coliforms are harmless to the human body, they make up the natural intestinal flora of man and are excreted abundantly, this group is used as an indicator of fecal contamination. Pathogenic organisms are the real causes of disease, such as Salmonella, Taenia, Schistosoma and Ascaris. These organisms are highly infectious and are responsible for thousands of deaths every year, especially in regions with poor health conditions. They are usually excreted by people with gastrointestinal diseases such as typhoid and paratyphoid fever, dysentery, diarrhea and cholera. As identifying these organisms is extremely difficult and time-consuming, the coliform group is used as an indicator of the presence of feces because they are more abundant, each person excretes around 100 to 400 billion coliforms per day, which makes them easier to detect (METCALF & EDDY, 2003). The discharge of effluents with concentrations of pathogenic organisms jeopardizes public health by transmitting diseases.

The filter membrane method was used, with a dilution of 10^{-3} for the affluent and 10^{-1} for the effluent. This method consists of filtering a certain volume of the sample, which will vary with its concentration, through a membrane that retains the coliforms. This membrane is placed over the culture medium contained in the plates and incubated for a period of 24 hours in an oven at 44.5°C.

5.3 STATISTICAL ANALYSIS

The data obtained for all the parameters were subjected to descriptive statistical analysis in order to better visualize their amplitude and variability. Arithmetic means, standard deviation, median, variance, coefficient of variation, maximum and minimum values and data counts were obtained for all parameters except for thermotolerant coliforms, for which only the geometric mean, median, maximum and minimum values and data counts were performed.

In order to verify the efficiency of the system in terms of removing the parameters analyzed, a study was carried out comparing the FAC effluent for the three phases, checking the influence of varying the TDH on the system. To this end, a single-factor analysis of variance (ANOVA) with a 95% confidence level (or 5% significance level) was applied. In this statistical study, the variability or variance between observations indicates whether the differences observed between groups are probably real or merely due to chance (WiTTE & WiTTE, 2005).

6. RESULTS AND DISCUSSION

This chapter presents and analyzes the data obtained during the experimental investigation. In order to meet the proposed objectives, the arithmetic means, standard deviations, coefficients of variation, medians and maximum and minimum values were determined for the effluents from the UASB reactor and the FAC during the experiment. The purpose of applying ANOVA (single factor) to the FAC effluent data was to check whether the variation in TDH between the three phases influenced the quality of the effluent. It was also applied to the uAsB effluent, even though the same operating conditions were maintained for this effluent, in order to check whether this effluent showed any statistical difference throughout the experiment and its interference with the effluent from the gypsum anaerobic filter.

The analysis of variance applied to the UASB reactor effluent resulted in no statistically significant difference between the three phases of the experiment in terms of organic matter and solids fractions, but for the other parameters, electrical conductivity, nutrients and thermotolerant coliforms, significant differences were observed, thus justifying a differentiated treatment in both the presentation and discussion of the results.

The results of the parameters analyzed are attached, organized in Tables, showing their respective descriptive statistics, including the analysis of variance applied to the UASB reactor effluent.

6.1 STATISTICAL TREATMENT

The study of the FAC effluent with different TDH was carried out by applying ANOVA to the experimental data, assuming a single factor (independent variable). Table 9 shows the results of the statistical treatment applied to the data obtained experimentally from the three groups according to the TDH applied to the FAC effluent. The analysis can be carried out by comparing the calculated and critical F values shown in Table 9. The following hypotheses were used in this analysis:

Null hypothesis - H0: no difference between the data groups

Alternative hypothesis - H1: there is a difference in at least one of the data groups.

When the calculated F is greater than the tabulated critical F, its value falls within the rejection region of the null hypothesis (H0), concluding that, at a significance level of 5%, there is a statistically significant difference between the three groups evaluated, thus adopting the alternative hypothesis (H1). However, if the opposite occurs, the calculated F value is less than or equal to the critical F, then it is considered that there is no statistically significant difference between the data obtained and analyzed, and the null hypothesis is accepted since the F value remains in the acceptance region.

Table 9. F values for the FAC effluent comparing the three phases.

Parameters	F cal.	Critical F	Region of acceptance	Rejection region
BOD5	1,356	3,315	X	
COD	2,457	3,315	X	
ST	0,422	3,315	X	
STV	1,693	3,315	X	
SST	2,969	3,315	X	
SSV	2,186	3,315	X	
Turbidity	3,373	3,402	X	
Electrical conductivity	4,259	3,315		X
Ammoniacal Nitrogen	11,207	3,244		X
Total Nitrogen	9,830	3,244		X
Organic Phosphorus	1,453	3,238	X	
Total Phosphorus	2,468	3,238	X	
Thermotolerant coliforms	5,307	3,327		X

As can be seen in Table 9, for all parameters except electrical conductivity, nitrogen and coliforms, the F value is in the region where the null hypothesis is accepted, and it is believed that there is no significant difference between the data obtained for the FAC effluent in the three phases of the experiment, under the conditions in which the experiment was carried out, concluding that the determining factor for obtaining better efficiency in the performance of the FAC is not related to varying the TDH; the analysis of variance carried out revealed that this was not the predominant hydraulic characteristic to characterize an increase in efficiency. However, for nitrogen and thermotolerant coliforms, the analysis of variance showed that there was a statistically significant difference for the FAC effluent in the three groups analyzed, in which each group has different TDH, 7, 9 and 10 days.

6.2 ORGANIC MATTER (BOD5and COD)

The most common way of assessing the organic matter content in sewage is through the determination of COD and BOD5, the latter in particular, as it directly measures the amount of biodegradable organic matter. These parameters are used as requirements to establish effluent quality standards by many environmental bodies.

Table 10 shows the average values for concentration, standard deviation and coefficient of variation for organic matter, obtained in each of the three phases of UASB effluent monitoring.

Table 10. Average BOD5 and COD values of UASB effluents.

1ª Phase	Average			
	BOD5		COD	
	UASB	FAC	UASB	FAC
Concentration (mg.L-1)	38,9	14	214	74
Standard deviation (mg.L-1)	2,6	4,1	42	20
Coefficient of variation (%)	6,7	29,4	20	27
Number of determinations	13	13	14	14
Stage 2	Average			
	BOD5		COD	
	UASB	FAC	UASB	FAC
Concentration ($mg.L^{-1}$)	36,0	12	253	71
Standard deviation ($mg.L^{-1}$)	4,0	3,3	44	32
Coefficient of variation (%)	11,2	27,6	17	46
Number of determinations	8	8	7	7
Stage 3	Average			
	BOD5		COD	
	UASB	FAC	UASB	FAC
Concentration ($mg.L^{-1}$)	40,1	12	228	56
Standard deviation ($mg.L^{-1}$)	4,9	3,1	70	19
Coefficient of variation (%)	12,3	26,2	31	33
Number of determinations	12	12	12	12

During the monitoring of the experimental system, the BOD5 of the effluent produced in the UASB reactor remained at an average of 39 mgO2/L in the 1st ª phase, maintaining an approximate average in the two consecutive phases, reaching a minimum value of 30 mgO2/L and a maximum value of 47 mgO2/L, both obtained during the 3rd ª phase of the experiment. With regard to COD, the same effluent maintained an average of 232 mg/L in the three phases, with values ranging from 124 to 387 mg/L of COD, both obtained during the 3rd phase, thus representing the highest coefficient of variation obtained in the experiment for this parameter, 31%.

Table 10 shows that after treatment in the filter, in the first phase, the final effluent had a low BOD5, maintaining an average of 14 mgO2/L and varying between 6 and 20 mgO2/L. In the 2nd and 3rd phases, there was a reduction in BOD5 values, but the FAC effluent in both phases maintained the same average value of 12 mgO2/L, with a variation in concentration between 7 and 17 mgO2/L. The values for COD were also observed, showing that in the first phase of the experiment, the FAC effluent maintained an average of 74 mg/L of COD, with minimum and maximum values of 39 and 100 mg/L, respectively. For the 2nd and 3rd phases, the minimum values obtained were 40 and 22 mg/L COD and the maximum values were 121 and 91 mg/L.

Figures 11 and 12 show the variations in BOD5 and COD, respectively, of the effluents from the UASB reactor and the FAC over the course of the experiment. It can be seen in the figures that the effluents have very different values, representing good removal; it can also be seen that the variation between the phases does not reflect on the behavior of the FAC effluent in both parameters, remaining practically within the same variation throughout the experiment.

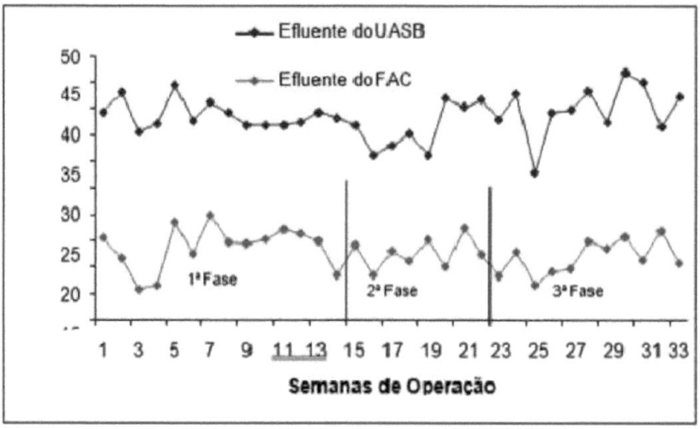

Figura 11. Performance of UASB and FAC effluents in terms of BOD5 during the experiment.

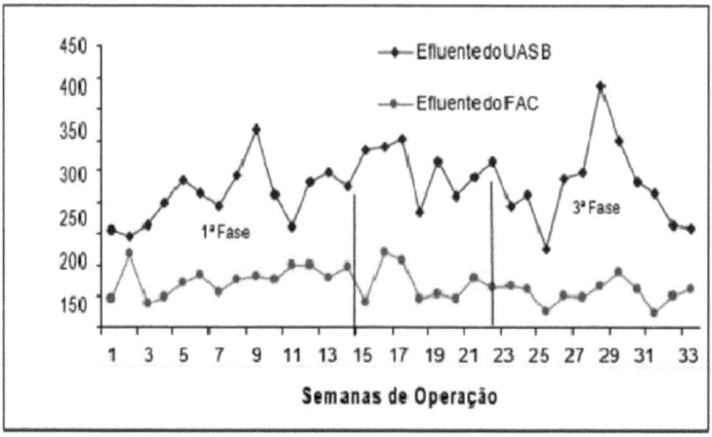

Figura 12. Performance of the effluents produced in the system in relation to COD in the three phases.

Figures 11 and 12 also show a regularity in the behavior of the effluent from the FAC, verifying a linearity in the data for both BOD5 and COD throughout the experiment; attesting to the fact that *there is* no difference in the filter effluent between the three phases, as statistically proven.

Kato et al. (1999) explains that the lower the flow rate applied, the lower the flow velocity and the longer the liquid will remain in contact with the biofilm. Initially, the biodegradable organic matter will be easily assimilated, but this assimilation will continually cease due to a lack of available organic matter. With the ideal flow, i.e. with the speed necessary for the organic matter to be assimilated without the liquid stagnating, the contact between the liquid and the biofilm will favor the absorption of the organic matter, allowing a greater quantity of sewage to receive treatment.

When the flow is maintained at an adequate speed, biofilm is produced uniformly over the entire length of the container. If the flow is slow, the initial area of the container will absorb all the assimilable organic matter, favoring rapid biofilm growth and causing this area to become clogged. With a faster flow, there will be enough organic matter to provide more uniform growth of the biofilm throughout the surface area of the FAC; taking care to maintain a speed that is not too fast, because then the biofilm will not have time to absorb and even if it passes through a large amount of biofilm, the fluid will come out with a high concentration of organic matter (CHARACKLIS & MARSHALL, 1989).

6.3 TOTAL SOLIDS

The averages of the statistical parameters for total solids (TS) and total volatile solids (TVS) values of the UASB and FAC effluents in the three phases of the experiment are shown in Table 11.

Table 11. Averages for total solids of the UASB and FAC effluents.

1ª Phase	**Average**			
	ST		STV	
	UASB	FAC	UASB	FAC
Concentration (mg.L^{-1}.)	722	611	200	127
Standard deviation (mg.L 1)	58	71	56	41
Coefficient of variation (%)	8	12	28	32
Number of determinations	14	14	14	14
Stage 2	**Average**			
	ST		STV	
	UASB	FAC	UASB	FAC
Concentration (mg.L^{-1},)	761	568	203	107

Standard deviation (mg.L 1)	46	64	40	62
Coefficient of variation (%)	6	11	20	58
Number of determinations	8	8	8	8
Stage 3	**Average**			
	ST		STV	
	UASB	FAC	UASB	FAC
Concentration (mg.L^{-1},)	667	587	162	99
Standard deviation (mg.L 1)	74	98	45	30
Coefficient of variation (%)	11	16,7	28	30
Number of determinations	12	12	12	12

The effluent produced in the UASB reactor had an average concentration of 717 mg/L in the three phases, with a minimum value of 512 mg/L obtained during phase 3ª and a maximum of 830 mg/L in phase 1, showing a low coefficient of variation (11%). The concentration for the same effluent in relation to STV was 200 mg/L for the two initial phases and 162 mg/L for the 3rd phase, representing an average of 188 mg/L for the three phases, with limit values between 73 and 344 mg/L, with a coefficient of variation of 28% in the 1st and 3rd phases, and 20% in the 2nd phase.

Table 11 shows that after treatment in the FAC, in the first phase, the effluent produced remained at an average of 611 mg/L for total solids, with limit values between 485 and 771 mg/L. Similar values were obtained in the consecutive phases, with no significant ST removal being observed for the FAC effluent during the three phases, maintaining an average concentration of around 600 mg/L. As for STV, the FAC effluent during phase 1ª remained at an average of 127 mg/L with limit values between 55 and 189 mg/L, with a coefficient of variation of 32%. Lower concentrations were obtained over the course of the experiment, with minimum values of 36 and 59 mg/L and maximum values of 193 and 172 mg/L for the 2nd and 3rd phases, respectively.

The behavior of ST was similar to that of STV, as can be seen in Figures 13 and 14. For both parameters, the values obtained in the effluents analyzed throughout the experiment remained close, with the FAC effluent remaining lower than the UASB effluent in most of the analyses carried out in the three phases, showing that even if small, there was removal. Figure 13 clearly shows that the two highest STV concentrations in the UASB effluent are found in the 1st phase and the lowest in the 3rd phase; however, for the FAC effluent, both the highest and lowest concentrations were obtained during the 2nd phase.

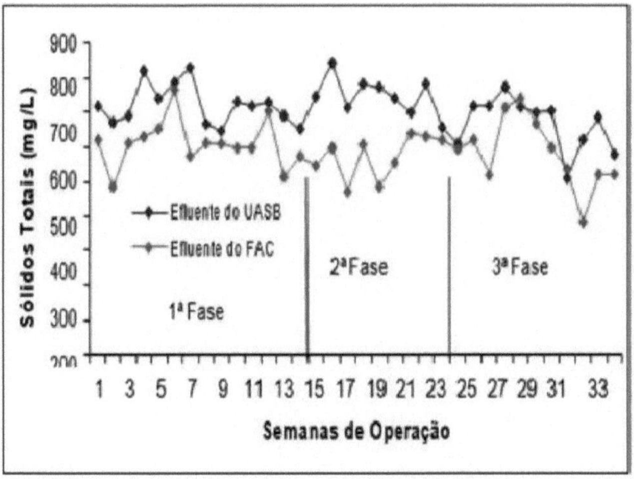

Figura 13. Behavior of UASB and FAC effluents in relation to ST during the experiment.

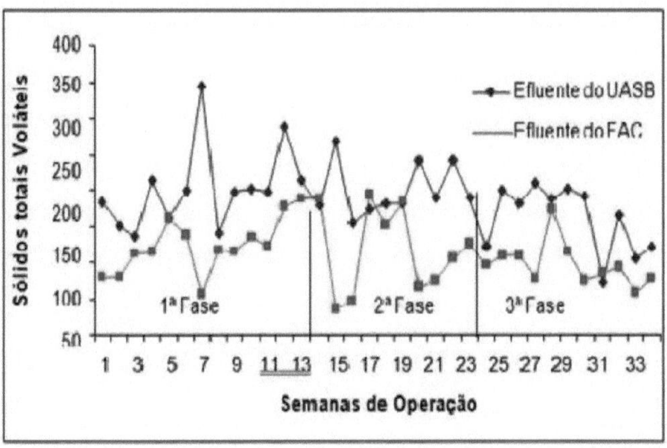

Figura 14. STV concentration for the UASB and FAC effluents during the experiment.

6.4 SUSPENDED SOLIDS

The average values for concentration, standard deviation and coefficient of variation for total suspended solids (TSS) and volatile suspended solids (VSS) of the UASB and FAC effluents in the three phases of the experiment can be seen in Table 12.

Table 12. Averages for suspended solids in the UASB and FAC effluents.

1≡ Phase	Average			
	SST		SSV	
	UASB	FAC	UASB	FAC
Concentration (mg.L^{-1})	58	9	48	7
Standard deviation (mg.L^{-1})	28	5	24	4
Coefficient of variation (%)	48	56	50	58
Number of determinations	14	14	14	14
2≡ Phase	**Average**			
	SST		SSV	
	UASB	FAC	UASB	FAC
Concentration (mg.L^{-1})	86	13	75	9
Standard deviation (mg.L^{-1})	27	7	27	4
Coefficient of variation (%)	32	49	36	44
Number of determinations	8	8	8	8
3≡ Phase	**Average**			
	SST		SSV	
	UASB	FAC	UASB	FAC
Concentration (mg.L^{-1})	70	10	55	7
Standard deviation (mg.L^{-1})	22	3	17	2
Coefficient of variation (%)	32	30	30	27
Number of determinations	12	12	12	12

For the three phases of the experiment, the effluent from the UASB reactor had an average concentration of 67 and 55 mg/L for TSS and VSS, respectively. With limit values between 27 and 127 mg/L for TSS, representing a coefficient of variation of 38%; and between 17 and 116 mg/L for SSV, resulting in a coefficient of variation of 50% throughout the experiment.

In phase 1^{a}, as shown in Table 12, the FAC effluent had an average TSS concentration of 9 mg/L with a coefficient of variation of 56%, the result of oscillations between 1 and 17 mg/L of this parameter. Although the TSS concentrations in the later phases were higher, with minimum values of 6 and 7 mg/L and maximum values of 23 and 15 mg/L for the 2nd and 3rd phases, respectively, the average concentrations of the FAC effluent in the three phases were close to 10 mg/L of TSS. Table 12 shows that in the three phases the FAC effluent values for SSV were similar, with an average concentration of just over 7 mg/L.

The behavior of the suspended solids in the effluents from the UASB reactor and the FAC can be seen in Figures 15 and 16. There is a similar behavior between the two parameters, TSS and VSS, with high variations in the UASB effluent in the three phases of the experiment, while the FAC effluent tends to decrease its variation throughout the phases.

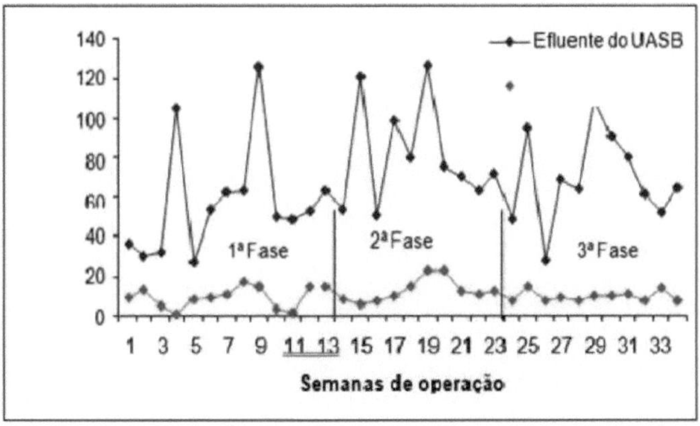

Figura 15. TSS values for UASB and FAC effluents during the experiment.

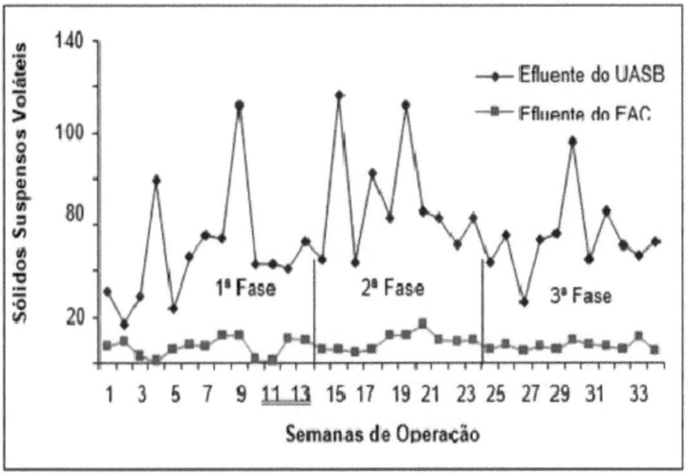

Figura 16. Figure 16. Performance of the UASB and FAC effluents in terms of SSV in the experiment.

6.5 TURBIDITY

Figure 17 shows the behavior of the turbidity values for the effluents from the UASB reactor and the Anaerobic Gum Filter during the period of operation of the experimental system.

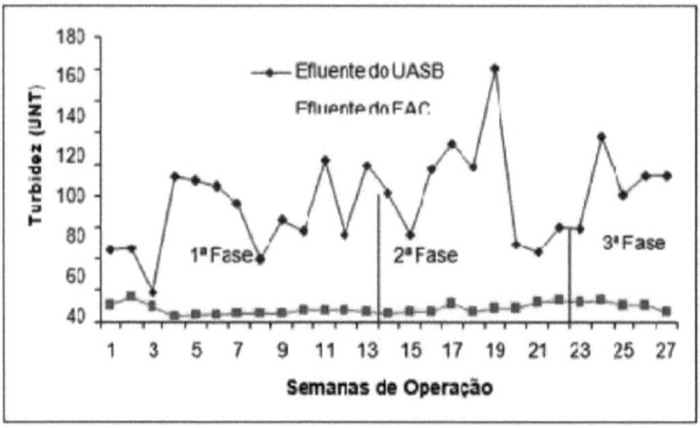

Figura 17. Turbidity behavior of UASB and FAC effluents.

It can be seen in Figure 17 that the effluent from the UASB reactor showed high variations during the experimental period, especially during phase 2ª where the maximum value was obtained (161 UNT - nephelometric turbidity unit), maintaining an average turbidity of 77 UNT in the three phases, with a coefficient of variation of 38%.

According to Figure 17, the effluent produced in the Anaerobic Filter showed low turbidity during the experiment, ranging from 4 to 16 UNT.

The analysis of variance showed no statistically significant difference in turbidity for either the UASB or FAC effluent. As can be seen in Figure 17, the UASB effluent shows large variations in the three phases that comprised the experiment, but the FAC effluent has close values between the three phases, ranging from 7 UNT in phase 1ª to 11 UNT in phase 3, with an average of 9 UNT.

However, the FAC effluent that showed the best results was during the 1st phase, with a value of 7 UNT, promoting an efficiency of 90%. Therefore, the FAC produces a well-clear effluent, which can be classified under CONAMA No. 357 of 2005 for class 1 water (CONAMA.2005).

6.6 ELECTRICAL CONDUCTIVITY

During phase 1ª the effluent from the UASB reactor had an average electrical conductivity of 1.50 $dS.m^{-1}$ with limit values of 1.4 to 1.6 $dS.m^{-1}$, representing a coefficient of variation of 3.6%. In both the 2nd and 3rd phases, the UASB effluent had a slightly lower average than in the 1st phase, at 1.4 dS.m^{-1}. However, in the 2nd phase the UASB effluent varied between 1.4 and 1.5 dS.m^{-1}, and in the 3rd phase it varied between 1.1 and 1.6 dS.m^{-1}.

Table 13 shows the averages of the statistical parameters for the conductivity values of the FAC effluent in the three phases.

As can be seen in Table 13, the two initial phases show similar average values, with a decrease in the last phase, since this is where the lowest conductivity values were obtained for the UASB effluent throughout the experiment and it even shows the highest coefficient of variation (9.5%). As can be seen, the electrical conductivity did not undergo any significant changes in terms of the treatment applied in the three phases, maintaining a value in the FAC effluent of around 1.4 dS.m^{-1} without any major variations.

Table 13. Average electrical conductivity values for the FAC effluent.

Statistical parameters	la Fase	Stage 2	Stage 3
Values (dS.m^{-1})	1,42	1,41	1,3
Standard deviation (dS.m^{-1})	0,06	0,11	0,15
Coefficient of variation (%)	4,1	7,9	11,6
Number of determinations	14	7	12

According to the results obtained, the FAC effluent can be used to irrigate crops, although certain precautions are necessary when using it. Electrical conductivity is directly related to the concentration of dissolved salts in the liquid, and a high concentration of these salts in the effluent used for irrigation impairs water absorption by the plants.

When it comes to crops that are classified as sensitive to salt tolerance, the FAC effluent has a concentration that is considered high. This was the case with the sesame crop grown in an experiment by Lima (2003), who used an effluent with a conductivity similar to the FAC effluent, and did not compromise it thanks to the characteristics of the soil used in the experiment (sandy soil). However, the effluent could be used to irrigate various other crops with greater tolerance than sesame, such as rice, cotton and soybeans, according to the classification presented by Gheyi et al. (1997).

6.7 NUTRIENTS

The averages obtained for the UASB and FAC effluents in terms of nitrogen and phosphorus for the three phases of the experiment are shown in Table 14.

Table 14. Average values of the nutrients analyzed in the system's effluents during the experimental period.

1ª Phase	Average							
	NTK		N-NH(L■		EN		P-PO4	
	UASB	FAC	UASB	FAC	UASB	FAC	UASB	FAC
Concentration (mg.L^{-1})	53,2	38,4	49	36,2	6,97	4,37	5,60	3,72
Standard deviation (mg.L	5,5	4,6	4,4	3,8	0,5	1,1	0,6	0,9
Coefficient of variation (%)	10,3	11,9	9	10,6	7,5	25,2	11,2	23,9
Stage 2	**Average**							
	NTK		N-NH$_L$'		EN		P-PO4	
	UASB	FAC	UASB	FAC	UASB	FAC	UASB	FAC
Concentration (mg.L^{-1})	54,7	38	47,1	34,4	6,73	4,45	5,87	4,02
Standard deviation (mg.L	3,0	4,6	3	6	0,6	0,6	0,4	0,7
Coefficient of variation (%)	5	12	7	17	8,4	14,4	6,3	16,6
Stage 3	**Average**							
	NTK		N-NH$_L$'		EN		P-PO4	
	UASB	FAC	UASB	FAC	UASB	FAC	UASB	FAC
Concentration (mg.L^{-1})	42,7	31,3	37,5	28,7	5,91	4,02	5,11	3,59
Standard deviation (mg.L	4,0	4	4	4	0,7	0,6	0,9	0,6
Coefficient of variation (%)	9	10	10	13	11,4	14,7	17,2	16,7

As shown in Table 14, the nutrient concentrations in the UASB reactor effluent remained higher than those in the FAC effluent throughout the experiment. In phases 1ª and 2ª, the UASB reactor effluent had an average concentration of over 50 mg/L for total nitrogen and close to this for ammonia. However, in the 3rd phase, lower concentrations were observed, around 40 mg/L for both total and ammoniacal nitrogen. This is reflected in the concentrations of the same nutrients for the FAC effluent, resulting in averages above 30 mg/L for nitrogen in the first two phases and below this in the 3rd phase for ammoniacal nitrogen. As for phosphorus, Table 14 shows similar concentrations in the three phases for both total and ammoniacal phosphorus. It can be seen that the effluent from the UASB reactor maintained average concentrations of 6 and 5 mg/L for total phosphorus and orthophosphate, respectively; and an average of 4 mg/L for phosphorus in both forms for the effluent from the Anaerobic Filter.

Figures 18, 19, 20 and 21 show the behavior of the UASB and FAC effluents for the two forms of nitrogen and phosphorus analyzed during the three phases of the experiment.

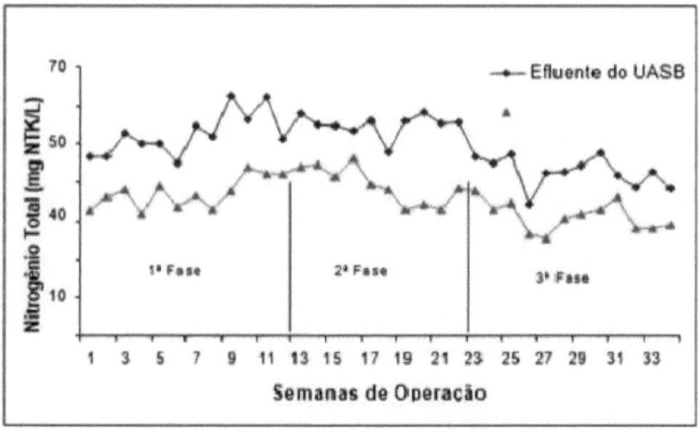

Figura 18. NTK values of the UASB and FAC effluents obtained in the three phases.

The effluent from the UASB reactor is characterized by high concentrations of nitrogen and phosphorus. The limit values obtained during the experiment for nitrogen were between 34.3 and 62.9 mg NTK.L^{-1} and 29.1 and 59.4 mg N-NH4+.L^{-1}, noting that the minimum concentrations were obtained in phase 3ª and the maximum in phase 1 for both forms of nitrogen; total phosphorus ranged from 4.6 to 7.9 mg P.L^{-1} and orthophosphate varied from 3.6 to 6.8 mg P-PO4 L^{-1}. Despite the lower concentrations, the FAC effluent showed very considerable values for nitrogen and phosphorus, ranging from 25.6 to 44.9 mg NTK.L^{-1} and 23.5 to 44.0 mg N-$NH4^{+}$ .L^{-1} for total and ammoniacal nitrogen, respectively; total phosphorus ranged from 2.3 to 6.2 mg P.L^{-1}, and orthophosphate ranged from 1.9 to 5.3 mg P-PO4 L^{-}; it can be seen that the limit values for phosphorus in both forms were obtained in phase 1ª, all these variations were the result of data collected during the three phases of the experiment.

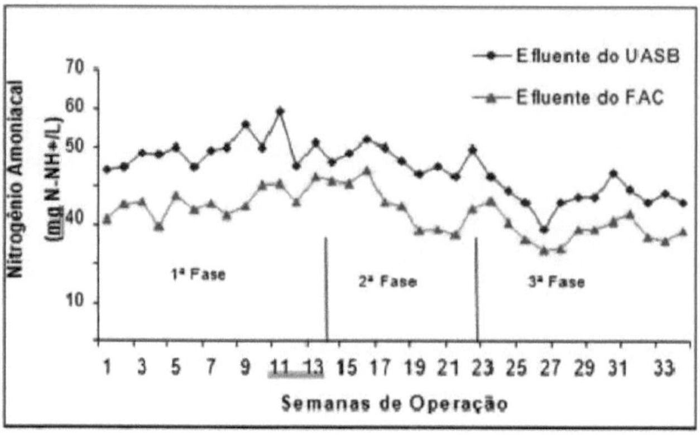

Figura 19. Behavior of the UASB and FAC effluents with regard to $N\text{-}NH^{+4}$ throughout the experiment.

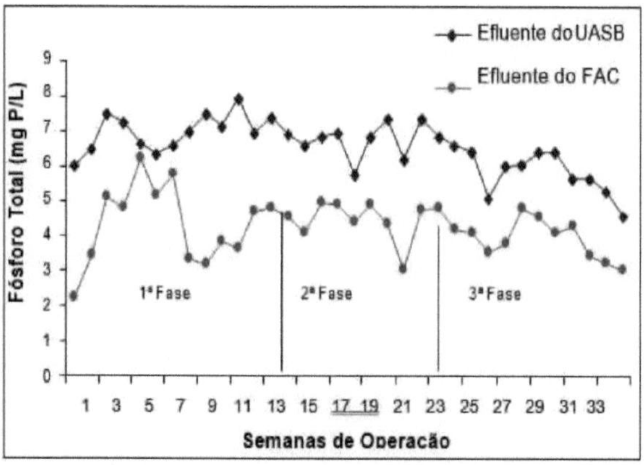

Figura 20. Total phosphorus concentrations for the system's effluents.

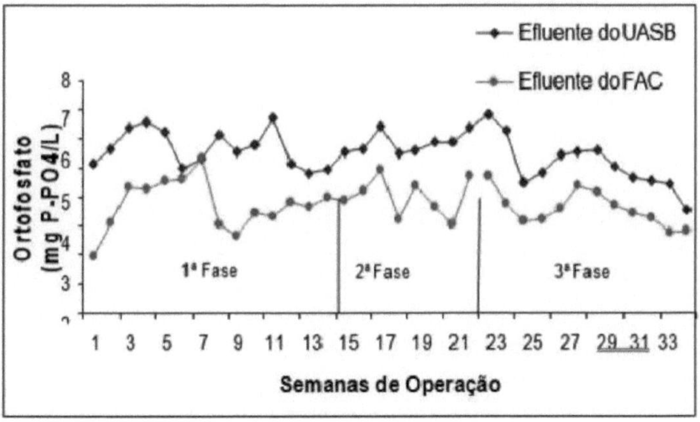

Figura 21. Behavior of UASB and FAC effluents in relation to orthophosphate throughout the experiment.

Figures 18, 19, 20 and 21 show a slight decrease in the concentration of nutrients in the effluent between the three phases, both for the UASB effluent and the Anaerobic Filter effluent.

The statistical analysis showed a significant difference in the UASB reactor effluent in the three phases for both nitrogen and phosphorus, in both forms, thus justifying the slope of the data. This statistical difference was reflected in the FAC effluent values for nitrogen, indicating a significant difference between the three phases. For this nutrient, the time factor is relevant in helping to improve the efficiency of the FAC. However, the same was not observed for phosphorus, since the analysis of variance showed that there was no significant difference in the FAC effluent between the phases.

6.8 THERMOTOLERANT COLIFORMS

For phase 1ª, a minimum value of 2.5 10^5 UFC/100mL and a maximum value of 8.65 10^6 UFC/100mL were obtained for the thermotolerant coliform group in tests carried out on the effluent from the UASB reactor, obtaining a geometric mean of 2.9 106 UFC/100mL. In the 2nd phase with the same effluent, a geometric mean of 8.13 106 UFC/100mL was obtained, with a minimum value of 3.3 106 UFC/100mL and a maximum value of 2.09 107 UFC/100mL for the thermotolerant coliform group. In the 3rd phase, the effluent from the UASB reactor showed a minimum value of 1.85 10^5 UFC/100mL and a maximum of 7.20 106 UFC/100mL in tests carried out for the same parameter, with a geometric mean of 1.23 10^6 UFC/100mL.

The results obtained from the tests carried out on the FAC effluents are shown in Table 15.

Table 15. Averages for fecal contamination indicators for FAC effluents in each phase.

Coliforms TermOtcleranles(TO4,1UFCyiOOmL)

Parameters	1ª Phase	? Phase	3ª Phase
Geometric mean	ose?	3,11	2.13
Màxime	1,80	111	8,75
Minimum	0,325	1,25	0,15

Throughout the experiment, the minimum value obtained in the tests carried out on the FAC effluent was 1.50 1 0^3 UFC/100mL and the maximum was 1.11 105 UFC/100mL, with a geometric mean of 1.56 10^4 UFC/100mL. As can be seen in Table 15, the lowest average was obtained during the 1st phase, with an order of magnitude of 10^3, and in the two consecutive phases the final effluent from the FAC remained with an order of magnitude of 10^4. The microbial removal efficiencies for the final effluent in the three phases were **99.700%, 99.713%, 98.216%** consecutively.

With regard to the sanitary quality of the final effluent produced in the system, Figure 22 shows the behavior of the UASB and FAC reactor effluents with regard to thermotolerant coliforms during 31 weeks of system monitoring.

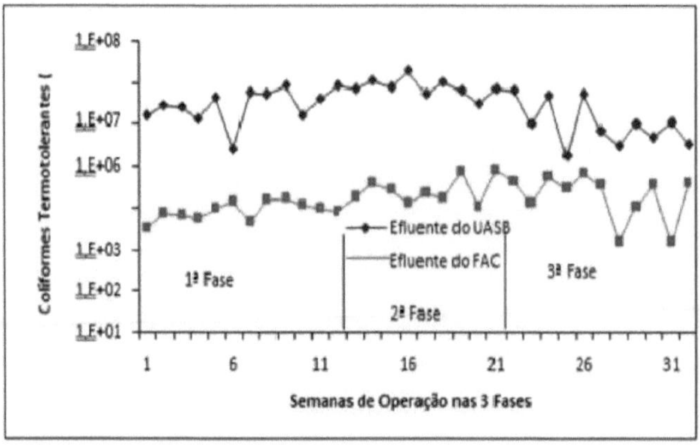

Figura 22. Behavior of Thermotolerant Coliforms in the three phases of the system.

The analysis of variance showed a statistically significant difference for the FAC effluent between the three phases, a result reflected by the UASB reactor effluent, which showed the same statistical result.

Bearing in mind that phase 1ª, within the conditions in which the experiment was carried out, was the one that showed the best results, a lower concentration of thermotolerant coliforms and greater removal, it is understood that for this parameter the increase in TDH impairs the efficiency of the FAC.

The FAC performed better in terms of thermotolerant coliform removal efficiency than the data obtained by Keller et al. (2004), using a UASB reactor and an aerated biofilter to treat domestic sewage. This showed a 1-log difference between the UASB effluent and the BF effluent, but in both experiments it was found that there was no satisfactory removal of this parameter.

Even though the FAC showed an efficiency of 99.700%, the final effluent does not meet the sanitary conditions established by the WHO (1989) for the irrigation of crops consumed raw, but for cereal crops and tree plants the effluent does not present any risk of contamination.

6.9 FAC EFFICIENCY

In order to check the performance of the system developed, FAC, comparing the three phases of the experiment, an analysis was made of the removal efficiency of the parameters analyzed. Table 16 shows the removal efficiency in terms of organic matter, COD and BOD5, solids fractions and nutrients, nitrogen and phosphorus.

The FAC performed well in terms of removing organic matter and suspended solids, as shown in Table 16.

Table 16. Removal efficiency (%) of FAC for organic matter, solids and nutrients in the three phases.

Phase	COD	BOD5	ST	STV	SST	SSV	NTK	N-NH4	EN	P-PO4
1ª	65	64	15	25	81	83	28	26	37	34
2ª	73	66	12	44	84	86	30	27	34	32
3ª	75	71	12	36	85	86	27	23	32	30

Table 16 shows that over the three phases for organic matter, the efficiency of the FAC tended to improve. Even so, this increase is not significant, as it was not enough to produce an effluent with low BOD5 concentrations (<10 mg O2/L). Furthermore, the increase in efficiency is not significant enough to justify increasing the TDH, as this will lead to an increase in operating costs.

It can therefore be concluded that for greater removal efficiency and better operation of the system for organic matter, within the study carried out, a TDH of 7 days can be used, with the effluent passing through the entire reactor, resulting in efficiencies of 65 and 64% for COD and BOD5, respectively.

This result proves to be satisfactory, as the effluent does not need to be retained for a long time in the FAC, presenting an effluent with a low concentration (14 mg O2/L) with very significant removal. The value of the final effluent is very close to that allowed by CONAMA in resolution 357 of 2005, which recommends a maximum limit of 10 mg O2/L for classifying this effluent in class 3.

Looking at efficiency, the FAC shows great effectiveness as a post-treatment system. Other experiments with biological filters applied in the post-treatment of anaerobic reactors, despite having a lower TDH and operating aerobically, obtained very similar efficiencies. Carvalho and Povinelli (2005), for example, with their submerged aerated biofilter operating for 6 hours, obtained average efficiencies of 75 and 78% for BOD and COD, respectively. Gonçalves and Araùjo (1998), also with a submerged aerated biofilter operating for 0.23 hours, had an efficiency of 72 and 56% for BOD and COD, respectively. The FAC can be compared to these systems with great advantages: it does not need electricity for aeration, since it works anaerobically, and consequently its sludge production is low, so there is no need to interrupt treatment for cleaning.

When compared to other anaerobic systems, the FAC shows better results. An example is the anaerobic biological filter filled with blast furnace slag, also used in the post-treatment of UASB effluent, which even in its best performance showed an average removal of 32% for COD and 52% for BOD5 (NASCIMENTO et al., 2000).

With regard to total solids, Table 20 shows that these were not significantly reduced by the treatment applied during the three phases of the experiment. The analysis of variance showed that no statistically significant difference was observed for this parameter, since the calculated F values remained lower than the critical F. As can be seen in Table 16, efficiency is not proportional to the increase in TDH; the best efficiency for ST, around 15%, is found in the effluent in phase 1ª, and for STV during phase 2, an efficiency of 44% was obtained.

However, the FAC achieved high efficiency values for suspended solids. Table 16 shows that the efficiency increases over the three phases, proving that in this case it is proportional to the variation in the TDH; obtaining values close to 90% efficiency in the removal of this parameter in the 3rd phase. However, the analysis of variance showed no significant difference for the FAC effluent in terms of suspended solids when comparing the effluent in the three phases, since, statistically, the increase in efficiency did not significantly influence the effluent concentrations.

The removal efficiency of the FAC for suspended solids was much higher than that obtained by Aisse et al. (2001) in their aerated biological filter, with around 50% for TSS, and also by Coraucci Filho et al. (2000) with four anaerobic filters filled with bamboo rings, removing 74% of TSS, and even when compared to other systems for polishing the effluent from the UASB reactor, such as the polishing ponds in the experiment by Mayer et al. (2001) with a range of efficiencies from 64 to 73% for TSS.

Even though the intervention of the TDH was statistically proven to improve the efficiency of the FAC , even if only in terms of nitrogen removal, no significant nitrogen removal was observed, as shown in Table 16. Bearing in mind that the anaerobic process is not characterized by the removal of nutrients, especially nitrogen and phosphorus, the treatment in the FAC provided a reduction of around 30% for both nitrogen and phosphorus, a removal attributed to the biological process of assimilation carried out in the biofilm by the microorganisms during the metabolism process.

6.10 ENVIRONMENTAL AND OPERATING CONDITIONS

To study the functioning of the FAC, the pH and alkalinity parameters were analyzed, bearing in mind that the pH of the medium is strictly important for the proper functioning of an anaerobic process, as is its buffering capacity.

6.10.1pH

Table 17 shows the pH variation for the UAsB and FAC effluents with their respective phases. It can be seen that the UAsB effluent was in a neutral range and after treatment in the FAC it became slightly alkaline, reaching a maximum pH of 8.2 in phases 1ª and 3ª. A lower pH was obtained during the 2nd phase, with a minimum value of 7.6 and a maximum of 8.0.

Table 17.pH of the three phases of the UAsB and FAC effluents.

Effluent	UAS		FA	
Limit values	Minimum	Maximum	Minim	Maximum
ia Phase	6,8	7,5	7,8	8,2
Stage 2	7,0	7,3	7,6	8,0
Stage 3	6,7	7,9	8,1	8,2

The behavior of the pH values obtained for the effluents from the UAsB reactor and the FAC during the three phases of the experiment can be seen in Figure 23.

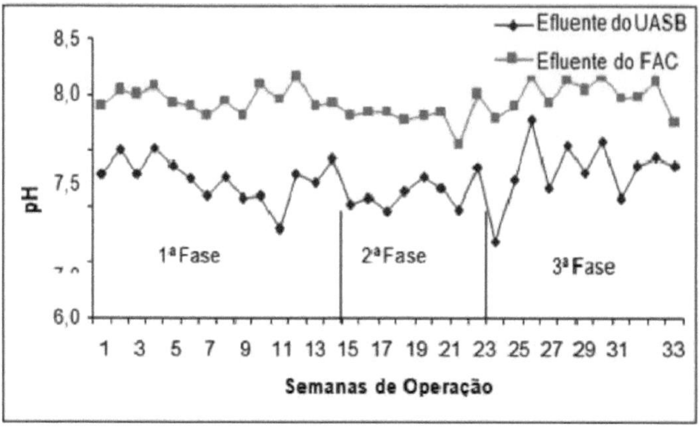

Figura 23. pH values of the effluents from the UASB reactor and the FAC during the experiment.

Figure 23 shows that the effluent from the UASB reactor was neutral throughout the experiment, rarely exceeding a pH of 7.5. After treatment in the FAC, the final effluent became slightly alkaline, always varying around pH 8.0. This increase in pH occurs during acid fermentation, where organic matter is converted into CO2 and the ammonification process takes place; however, this pH does not compromise the loss of ammoniacal nitrogen through ammonia volatilization (VAN HAANDEL & LETTINGA, 1994).

$$NH_4^+ \longrightarrow NH_3 + H^+ \qquad (10)$$

The characteristic of the FAC effluent in maintaining an increase in pH can be attributed to a characteristic of this type of treatment, since the same occurred with the four anaerobic filters monitored by Coelho (2001), two treating UASB reactor effluent and two septic tank effluent, with pH varying between 7.4 and 8.3.

6.10.2 Alkalinity

Table 18 shows the average values, standard deviation and coefficient of variation for each phase of the experiment for the UASB reactor and FAC effluents in terms of alkalinity. It can be seen that the effluent from the UASB reactor has a total alkalinity (TA) of over 300 mg CaCO3/L, exceeding 600 mg CaCO3/L in phase 2[a]. For the FAC effluent, the total alkalinity in the three phases was around 300 mg CaCO3/L. The averages for alkalinity to bicarbonate (AB) found during the experiment were between 300 and 340 mg CaCO3/L for the UASB effluent, and for the FAC between 260 and 290 mg CaCO3/L. As for volatile fatty acids (VFA), the UASB effluent showed an average value ranging from 60 mg Hac/L for the 1st and 3rd phases to 100 mg Hac/L for the 2nd phase, while the FAC effluent showed averages of less than 11 mg Hac/L in the three phases.

Table 18. Average alkalinity values in the three phases of the experiment.

ia Phase	UASB			FAC		
	AT	AB	AGV	AT	AB	AGV
Averages (mg/L)	380,7	340,1	69,8	309,9	293,1	11,4
Standard deviation (mg/L)	24,1	33,3	30,3	22,3	20,5	4,7
Coefficient of variation (%)	6,3	9,8	43,4	7,2	7,0	41,2
Stage 2	UASB			FAC		
	AT	AB	AGV	AT	AB	AGV
Averages (mg/L)	631,8	314,5	108,8	305,7	286,2	28,0
Standard deviation (mg/L)	19,6	31,2	23,5	20,3	20,2	14,9
Coefficient of variation (%)	5,4	9,9	21,6	6,6	7,1	53,2
Stage 3	UASB			FAC		
	AT	AB	AGV	AT	AB	AGV
Averages (mg/L)	341,9	302,3	65,4	279,1	266,5	15
Standard deviation (mg/L)	27,4	30,3	24,5	30,2	29,3	4,2
Coefficient of variation (%)	8,0	9,9	37,5	10,8	11,0	28,2

Although total alkalinity data was obtained, the results will only be analyzed and discussed in terms of bicarbonate alkalinity and the concentration of volatile fatty acids, as these are the parameters of interest for the stability of anaerobic reactors.

The same behavior was observed in all three phases for AB and AGV, a reduction in both parameters. When AGV *is* consumed, it is immediately assumed that it is converted into methane and, therefore, this would result in the production of alkalinity.

6.11 SUPPORT MATERIAL

The Chicana Anaerobic Filter (CAF) was filled with polyethylene terephthalate (PET) bottles. After various tests, 96% of the voids formed by the FAC support material were obtained. This material had a greater volume of voids than the materials used by Andrade Neto et al. (1997) in their comparative study. They used commercial and No. 4 gravel, sand and blast furnace slag, but the material with the best volume was 89% using cut conduit rings. A volume of voids similar to that obtained with PET bottles was the material used by Rolo and Além Sobrinho (2004) in their Biological Filter (97%).

The result obtained for the specific surface area of the support material was 160m$^{(2)}$/m^3, guaranteeing a surface application rate of 0.169 gDQO/m^3.day. This rate is considered relatively low when compared to the experiment carried out on an Aerated Biological Filter which worked with 14 gDQO/m^3.day (RÔLO & ALÉM SOBRINHO, 2004).

6.12 SLUDGE PRODUCTION

The behavior of total solids in each chamber can be seen in Figure 24. The first chamber had a ST concentration of 92 g/L, the second 91 g/L and the third 83 g/L. The STV analyses resulted in concentrations of 52, 51 and 46 g/L for the chambers, consecutively.

During the ten months of operation, the presence of sludge was detected in the initial period, relative to phase 1ª, revealing a low production of sludge generated by the FAC. This is due to the system's initial period of operation, when the biofilm had not yet formed and the organic matter, unable to adhere to the support material, tended to settle at the bottom of the FAC. Looking at Figure 24, you can see that there is a decrease in the concentration of solids as you progress through the chambers (C1>C2>C3), because at this point the FAC is only working as a filter, retaining the organic matter through the support material.

When the biofilm starts to form and grow, the organic matter starts to adhere to the support material and is digested by the microorganisms. Furthermore, the TRC is significantly high when it comes to biofilm, as observed by Reichert and Wanner (1997) through various simulations. Avoiding the deposit of sludge at the bottom of the FAC, as evidenced by the absence of sludge in the other two disposals; which does not mean that the system does not produce sludge, as the biomass remains adhered to the support material or retained in its interstices.

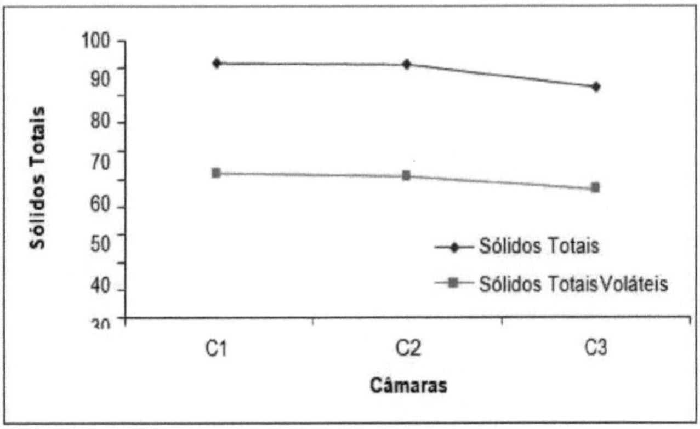

Figura 24. Total solids concentration of the sludge in each chamber.

In order to implement the alternative presented in this work for the treatment of sewage in small communities, a UASB reactor followed by a chicane anaerobic filter (FAC), some estimates were made based on the results obtained experimentally. Assuming that in a small and/or low-income community, each inhabitant would consume an average of 50 L daily. A UASB reactor with the same characteristics as the one used in this study, treating 20 m^3/day, would serve a population of around 400 inhabitants.

As we have seen, the effluent from the UASB would need to be polished before it could be used or even discharged into a body of water. A filter with sufficient capacity to treat all the UASB effluent on a daily basis would have to have a volume of 140 m3, with the filter operating at a TDH of 7 days, as seen in this work to be the most feasible.

For this size, adopting a depth of 1 m, the FAC would occupy an area of 140 m^2. For a geometry similar to the one used in the experiment, rectangular, we suggest the dimensions shown in Figure 25; we also suggest implanting 5 baffles in the FAC, with 1ª located 2 m from the introduction of the affluent, and the others every 1 m. This suggestion is based on the need for a larger area in the F chamber of the filter due to the greater concentration of solids.

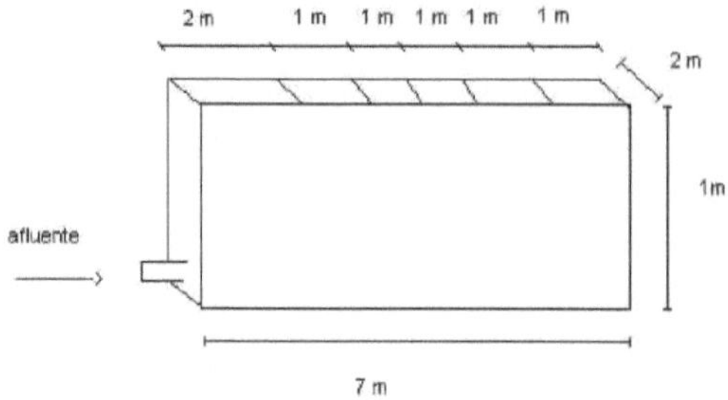

Figura 25. FAC dimensions for a population of 400 inhabitants.

- Statistically, there were no significant differences, at the 5% level, in relation to organic matter and suspended solids in the effluent produced in the Anaerobic Filter, which was subjected to the three different times of hydraulic detention; however, there was a significant difference in the removal of Thermotolerant Coliforms.

- The choice of phase 1ª, a TDH of 7 days, makes it possible to treat a greater quantity of sewage in a shorter space of time.

- The removal efficiency of total and volatile suspended solids in the Chicana Anaerobic Filter remained at an average of 85 and 86%, which is considerable considering that the filter is an effluent polishing system.

- The effluent produced by the Filter had a low organic load (BOD5 12mg. L^{-1}) and a high concentration of macronutrients, with a pH ranging from 7.3 to 8.2, so it can be used to irrigate a variety of crops.

- The Gum Anaerobic Filter used to post-treat the effluent from the UASB reactor produced effluent with a concentration of thermotolerant Coliforms in the order of 103 to 104 CFU. 100mL^{-1}, effluent with this characteristic can only be used for fertigation of forage crops and crops not consumed raw.

- The filling material used proved to be a viable alternative, as it provided a high volume of empties, thereby reducing the risk of clogging, and low implementation costs as it is a disposable material. PET bottles also have very favorable characteristics: they are light, resistant and inert.

- The Chicana Anaerobic Filter produced little sludge. The high volume of voids allowed for a large aggregation of organic matter, favoring its digestion and helping to reduce the risk of clogging.

- In order to treat sewage for a low-income population of around 400 inhabitants, it is necessary to install a system consisting of a $5m^3$ UASB reactor and a $140m^3$ FAC that can treat 20m3 per day.

7. REFERENCES

ABNT - ASSOCIAÇÃO BRASILEIRA DE NORMAS TÉCNICAS. NBR 7229 - Construction and Installation of Septic Tanks and Disposal of Final Effluents, 1982, 37p.

AISSE, Miguel Mansur, JÜRGENSEN, Décie, ALÉM SOBRINHO, Pedro, Avaliação do Reator UASB e Filtro Biológico para Tratamento de Esgotos Sanititários in: **Pós- tratamento de Efluentes de Reatores Anaeróbios**: Coletânea de Trabalhos Técnicos, v. 2, Belo Horizonte-MG, Coordenador C. A. L. Chernicharo, 2001. CD-Rom.

ALMEIDA, José Berivaldo Sales, **Removal of Heavy Metals**: Impact on Health and the Environment, Campina Grande-PB, November-2004, Monograph.

ANDRADE NETO, Cicero Onofre, **Simple Systems for Sewage Treatment:** Brazilian Experience, Rio de Janeiro: ABES, 1997, 45p. , 54p.

ANDRADE NETO, Cicero O.; CAMPOS, José R.; ALÉM SOBRINHO, Pedro; CHERNICHARO, Carlos A. L.; NOUR, Edson A., Filtro Anaeròbio in: **Tratamento de Esgotos Sanititários por Processo Anaeròbio e Disposiçao Controlada no Solo**, Rio de Janeiro-RJ, Coordenador J. R. Campos, projeto PROSAB, 1999, 140p., 145p.

ANDRADE NETO, Cicero O.; MELO, Hênio N.S.; FILHO, Manoel L., **Filtros Anaeróbios com Fluxo Ascendente e Fluxo Descendente** in: Pós-tratamento de Efluentes de Reatores Anaeróbios: Coletânea de Trabalhos Técnicos, Vol 2, Belo Horizonte-MG, Coordenador C. A. L. Chernicharo, 2001, 185p. CD Rom.

ANDREOTTOLA, G., FOLADORI, P., NARDELLI, P., DENICOLO, A., **Treatment of Winery Wastewater in a Full-scale Fixed Bed Biofilm Reactor**, Water Science and Technology v. 51, n. 1, 2005, p. 71-79.

APHA. AWWA. WPCF. **Standard methods for the examination of water and wastewater**. 15 ed. Washington, D.. American Public Health Association. American Water Works Association, Water Pollution control Federation, 1995, 1134p.

BISHOP, P. Biofilm Structure and Kinectics. **Water Science and Technology**, v.36 n.1 pp 287-294, 1997.

BRUSEKE, Franz Josef. The Problem of Sustainable Development in: **Development and Nature:** Studies for a Sustainable Society. Organizer: Clóvis Cavalcanti, 3ª edition, Sao

Paulo: Cortez, 2001.

CAMPOS, José Roberto (Coordinator). **Post-treatment of Anaerobic Reactor Effluents,** Belo Horizonte, Projeto PROSAB, 2001, 20 p.

CARVALHO, O.J.; POVINELLI, J. Submerged aerobic biofilter used in the post-treatment of compartmentalized anaerobic reactor effluent. In: **Brazilian** Congress of **Sanitary and Environmental Engineering**, 23, 2005, CD Rom.

CHARACKLIS, William G., MARSHALL, Kevin C. **Biofilms**, New York, Chichester: A wiley-Interscience, p. 796, 1989.

CHERNICHARO, Carlos Augusto de Lemos, **Principios do Tratamento Biològico** de **Aguas Residuárias**: Reatores Anaeróbios, volume 5, Belo Horizonte, UFMG, 1997, 154p., 138p. , 51p.

CHERNICHARO, C. A. L. (Coordinator). **Post-Treatment of Anaerobic Reactor Effluent.** Belo Horizonte (MG). PROSAB Project, 2001, 20p., 173p., 254p. 492p.

COELHO, André Luiz da Silva Salgado, **Estudo Comparativo entre Sistemas de Tratamento Unifamiliar de Esgoto com Pós-tratamento ou Disposiçâ final no Solo,** Campina Grande-PB, 2001, Dissertaçao de Mestrado, p.110.

CONAMA. NATIONAL ENVIRONMENT COUNCIL, No. 357, of March 17, 2005.

EIGER, Sérgio, Autodepuraçao dos Cursos de Agua in: **Reuso de Agua**, Pedro C. S. Mancuso e Hilton F. Santos, editors; Barueri, SP: Manole, p. 233-259, 2003.

FARABEGoLi, G., CARUCCi, A., GANDoLFo, G., RoLLE, E., VioTTi, P..**Experimental Study on Carbon Removal in Biological Aerated Filters**, Water Science and Technology Vol 48, n° 11-12, 2003, p. 235-242.

FoRESTi, Eugenio, FLoRÊNCio, Lourdinha, VAN HAANDEL, Adrianus, ZAiAT, Marcelo, CAVALCANTi, Paula F. F. Fundamentos do Tratamento Anaeròbio in: **Tratamento** de **Esgotos Sanitàrios por Processo Anaeròbio e Disposiçâo Controlada no Solo**, Rio de Janeiro-RJ, Coordenador J. R. Campos, projeto PRoSAB, 1999, p.29.

GHEYi, H. Y.; QUEiRoZ, J. E.; MEDEiRoS, J. F. Manejo e controle da salinidade na agricultura irrigada in: **Congresso Brasileiro de Engenharia Agricola**, 36. Campina Grande, 1997, 15p.

GONÇALVES, R. F., ARAÙJO, V. L. **Secondary treatment of sanitary sewage through the association in series of UASB reactors and submerged aerated biofilters**. Engenharia, Ciência e Tecnologia, Espirito Santo, March/April 3, p.14-19, 1998.

HO, G. **Small Water and Wastewater Systems: Pathways to Sustainable Development?Water Science and Technology v. 48, n. 11-12, 2003, p. 7-14**

IBGE. INSTITUTO BRASILEIRO DE GEOGRAFIA E ESTATISTICA, Pesquisa Nacional de Saneamento Bàsico, Rio de Janeiro: IBGE, 2000a, 397p.

IBGE. National Population Survey by Region. Resident Population, by Sex and Household Situation, Resident Population Aged 10 and Over, Total, Literate and Literacy Rate, by Municipality, 2000b. Available at< http://www.ibge.gov.br> Accessed on 15/02/2006.

IBGE. Municipal Social Indicators: Proportion of permanent private households, by type of sanitation, according to Federation Units and population size classes of municipalities - Brazil - 2000c. Available at< http://www.ibge.gov.br/home/estatistica/populacao/indicadores social municipais /tabela16c.shtm > Accessed on 28/02/2006.

IMHOFF, Karl and Klaus R., **Manual of Wastewater Treatment,** 1998, 1 p.

JANNING, K. F. MESTERTON, K, HARREMOËS, P., Hydrolysis and Degradation of Filtrated Organic Particulates in a Biofilm reactor under Anoxic and Aerobic Conditions in: **Water Science and Technology,** v. 36, n.1, p.279-286, 1997.

JORDÂO, Eduardo Pacheco; PESSÔA, Constantino Arruda, **Tratamento de Esgotos Domésticos,** 3rd edition, Rio de Janeiro: ABES, 1995.

KATO, Mario T., ANDRADE NETO, Cicero O., CHERNICHARO, Carlos A.L., FORESTI,Eugenio, CYBIS, Luis F. Configurations of Anaerobic Reactors in: **Treatment of Sanitary Sewage by Anaerobic Process and Controlled Disposal in Soil**, Rio de Janeiro-RJ, Coordinator J. R. Campos, PROSAB project, 1999, p.53.

KELLER, R. , PASSAMANI-FRANCA R. F. , PASSAMANI, F. , VAZ, L. , CASSINI, S. T., SHERRER, N. , RUBIM, K. , SANT'ANA, T. D. , GONÇALVES, R. F. Pathogen Removal efficiency from UASB+BF Effluent Using Conventional and UV post-treatment Systems,**Water Science and Technology** v. 50, n. 1, p. 16, 2004.

LAZAROVA, V. MANEM, **J.** Biofilm characterization and activity analysis in water and wastewater treatment in: **Water Science and Technology**, v. 29, n.10, 1995.

LEFF, Henrique. **Saber Ambiental:** sustentabilidade, complexidade e poder, 2ª ediçao, Petrópolis, RJ: Vozes, 2001.

LEONARDI, Maria Lùcia Azevedo. A Sociedade Global e a Questao Ambiental in: **Desenvolvimento e Natureza:** Estudos para uma Sociedade Sustentàvel. Clóvis Cavalcanti, 3ª ed., Sao Paulo: Cortez, 2001, 195p.

LIMA, Eduardo Pedroza da Cunha. **Use of Tracer for Hydrodynamic Assessment of Polishing Lagoon**. Final report. PIBIC/UEPB. CAMPINA Grande, 2001.

LIMA, Eduardo Pedroza da Cunha, **Treatment of Sanitary Sewage and its Use in Fertigation of Sesame**: A Proposal for Family Farming, 2003, 160p. Master's dissertation.

LOBO, Luiz sp. **Saneamento Bàsico:** Em Busca da Universalizaçao. Brasilia, Ed. do autor, 2003, p. 228

LOWY, Michael. From Marx to Ecosocialism in: **Post-Neoliberalism II:** What State for What Democracy? Organizers: Emir Sader and Pablo Gentili, Petrópolis, RJ: Vozes, 1999.

MAYER, Maria das Graças Ribeiro, MOREIRA, Eudes Alves, CAVALCANTI, Paula Frassinetti Feitosa, VAN HAANDEL Adrianus Polishing Lagoons for the Post-treatment of Digested Sewage - part 1: Removal of Organic Matter and Suspended Solids in: in: **Post-treatment of Anaerobic Reactor Effluents**: Collection of Technical Papers, v. 2, Belo Horizonte-MG, Coordinator C. A. L. Chernicharo, 2001. CD Rom.

METCALF & EDDY, **Wastewater Engineering:** treatment, disposal, reuse. 4th edition McGRAW-HILL. 2003, 11p.,455p., 276p. 1819p

MOLETTA, R. **Winery and distillery wastewater treatment by Anaerobic Digestion, Water Science and Technology**, v 51, n. 1 2005, p 137-144.

NASCIMENTO, Mauro C.P., CHERNICHARO, Carlos A.L., BEJAR, Deneb O., **Filtros Biológicos Aplicados ao Pós-tratamento** de **Efluentes** de **Reatores UASB,** in: Pós-tratamento de Efluentes de Reatores Anaeróbios: Coletânea de Trabalhos Técnicos, volume1, Projeto PROSAB, coordenador C. A. L. Chernicharo, Belo Horizonte-MG, 2000.

PELIZZOLI, M. L. **Correntes da Ética Ambiental,** Petrópolis, RJ: Vozes, 2002.

PNAD. NATIONAL HOUSEHOLD SAMPLE SURVEY. Permanent Private Households, by Household Monthly Income Classes, According to Household Situation and Some

Characteristics, 2003, recorded on: 18/01/2005. Available at < http: wwww.ibge.gov.br>

PONTING, Clive. **A Green History of the World,** Rio de Janeiro: Civilizaçâo Brasileira, 1995.

RANDALL, C.W. Changing Needs for Appropriate excreta disposal and small Wastewater Treatment Methodologies or The Future Technology of small Wastewater treatment systems, **Water Science and Technology**, v. 48 n. 1, , p. 16, 2003.

REiCHERT, Peter, WANNER, oskar, Movement of solids in Biofilms: significance of Liquid Phase Transport in: **Water Science and Technology,** v.36, n. 1, p. 321-328, 1997.

RÔLO, Màrcia Cristina, ALÉM SOBRINHO, Pedro **Post-treatment of UASB Reactor Effluent Using an Aerated Submerged Biological Filter with Alternative Filling**, XI SILUBESA, 2004.

SANTOS, Hilton Felicio, MANCUSO, Pedro Caetano Sanches, Participaçao Comunitària e Aceitabilidade da **Agua de Reuso** in: **Reuso de Agua,** Pedro C. S. Mancuso e Hilton F. Santos, editors; Barueri, SP: Manole, p. 470-477, 2003.

SAUNDERS, Robert J. ,WARFORD, Jeremy J. **Water Supply in Small Communities:** Aspectos Econômicos e Politicos nos Paises em Desenvolvimento, transl. Vera Lucia Mixtro Chama, 2ª ed. Rio de Janeiro:ABES; 2002, 62-79p.

SILVA, Antônio C.C. **Pacto Contra a Exclusao**, Bio: revista brasileira de saneamento e meio ambiente, ABES, ano XIII n. 28, Outubro - Dezembro 2003, 12 p.

VAN HAANDEL, Adrianus C.; LETTINGA, Gatze, **Tratamento Anaeróbio de Esgotos:**Um manual para regioes de clima quente, Campina Grande, Epgraf, 1994, 21 p. II-22p.

VAN HAANDEL A.; MARAIS G. **O Comportamento do Sistema de Lodo Ativado:**Teoria e Aplicaçoes para Projeto e Operaçao, Campina Grande-PB, Epgraf, 1999, 279p., 18p.

VAZ, Silvânia R. **A Iniciativa Privada no Saneamento Bàsico Brasileiro**, Gerenciamento Ambiental, Editora Ltda BJMoura, Sao Paulo, n. 22, November - December 2002, 15 p.

VON SPERLING, Marcos, **Introduçâo à qualidade das águas e ao tratamento de esgotos,** vol. 1, 2nd revised edition, Belo Horizonte, 1996a, 93 p.

VON SPERLING, Marcos, **Principios Basicos do Tratamento de Esgotos,** vol. 2, Belo

Horizonte, 1996b, 37 p.,17p

WHO. Health guidelines for the use of wastewater in agriculture and aquaculture, Technical Report Series, n. 778. Geneva, Switzerland: World Health Organization, 1989.

WITTE, R., WITTE, S, J. **Estatistica**. Translated by Tereza Cristina Padilha de Souza, LTC-Rio de Janeiro, 486p. 2005.

MIX
Papier aus verantwortungsvollen Quellen
Paper from responsible sources
FSC® C105338

Printed by Books on Demand GmbH, Norderstedt / Germany